SIMPLIFIED DIFFERENTIAL EQUATION

CHAPTER 1: INTRODUCTION: DEFINITION

1.7 SUPPLEMENTARY PROBLEMS

1. Determine the order, degree, and the type of the following differential equations:

a) $(x + y)dx + (3x^2 - 1)dy = 0$

Order = 1	Degree = 1	Type = Ordinary

b) $x\frac{d^2y}{dt^2} - y\frac{d^2x}{dt^2} = k$

Order = 2	Degree = 1	Type = Ordinary

c) $\frac{a^2v}{ax^2} + \frac{a^2w}{ax^2} = 0$

Order = 2	Degree = 1	Type = Partial

d) $x\left(\frac{d^2y}{dx^2}\right)^3 + \left(\frac{dy}{dx}\right)^4 - y = 0$

Order = 2	Degree = 3	Type = Ordinary

e) $y''' + 4y' + 3y = x$

Order = 3	Degree = 1	Type = Ordinary

2. Prove that each equation is a solution of the given differential equation:

a) $y = x^2 + 4x; x\frac{dy}{dx} = x^2 + y$

$y = x^2 + 4x$ $; x\frac{dy}{dx} = x^2 + y$

$\frac{dy}{dx} = 2x + 4$ $x(2x + 4) = x^2 + (x^2 + 4x)$

$2x^2 + 4x - 2x^2 - 4x = 0$

$\boxed{0 = 0}$

b) $y = A\sin 5x + B\cos 5x; \frac{d^2y}{x^2} + 25y = 0$

$y = A\sin 5x + B\cos 5x$

$\frac{dy}{dx} = A\cos 5x\ (5) - B\sin 5x\ (5)$ $; \frac{d^2y}{dx^2} + 25y = 0$

$\quad = -25\ (A\sin 5x + B\cos 5x)$ $-25y + 25y = 0$

$\frac{dy}{dx} = -25y$ $\boxed{0 = 0}$

c) $y = (x+1)e^{-x}$; $y' + y - e^{-x} = 0$

$$y = (x+1)\,e^{-x}$$

$$y' = (x+1) - e^{-x} + (e^{-x})1 + 0$$

$$= -xe^{-x} + (-e^{-x}) + e^{-x} + 0$$

$$= -xe^{-x}$$

$;y' + y - e^{-x} = 0$

$-xe^{-x} + (x+1)e^{-x} - e^{-x} =$

$-xe^{-x} + xe^{-x} + e^{-x} - e^{-x} =$

$\boxed{0 = 0}$

d) $lny = C_1 e^x + C_2 e^{-x}$; $yy'' - (y')^2 = y^2 lny$

$$\frac{y'}{y} = C_1 e^x + C_2 e^{-x}(-1)$$

$$\frac{yy'' - y'y'}{y^2} = C_1 e^x + C_2 e^{-x}(-1)$$

$$yy'' - y'^2 = y^2(C_1 e^x + C_2 e^{-x})$$

$; yy'' - (y')^2 = y^2\,lny$

$y^2(C_1 e^x + C_2 e^{-x}) = y^2\,lny$

$y^2\,lny - y^2\,lny = 0$

$\boxed{0 = 0}$

e) $y^{-3} = x^3(3e^x + C)$; $xy' + y + x^4 y^4 e^x = 0$

$$y^{-3} = x^3(3e^x + C)$$

$$\frac{y^{-3}}{x^{-3}} = (3e^x + C)$$

$$y^{-3}x^{-3} = (3e^x + C)$$

$$y^{-3}(-3x^{-4}) + x^{-3}(-3y^{-4}) = 3e^x$$

$$-x^{-4}y^{-3} - y^{-4}x^{-3}y' = e^x$$

$$\frac{y}{y}\left[-\frac{1}{x^4 y^3} - \frac{y'}{y^4 x^3}\right]\frac{x}{x} = e^x$$

$$\frac{-y - y'}{x^4 y^4} = e^x$$

$$-y - y'x = e^x(x^4 y^4)$$

$$e^x x^4 y^4 + y + y'x = 0$$

$; xy' + y + x^4 y^4 e^x = 0$

$\boxed{0 = 0}$

CHAPTER 2: SEPARATION OF VARIABLES

2.4 SUPPLEMENTARY PROBLEMS

Find the complete or particular solutions of the following differential equations:

1. $\dfrac{dy}{dx} = \dfrac{4x+xy^2}{y-x^2y}$

$\left[\dfrac{dy}{dx} = \dfrac{4x+xy^2}{y-x^2y}\right]$

$\left[\dfrac{dy}{dx} = \dfrac{x(4+y^2)}{y(1-x^2)}\right]\dfrac{dx}{x\,(4+y^2)}$

$\left[\dfrac{dy}{x(4+y^2)} = \dfrac{dx}{y(1-x^2)}\right]xy$

$\dfrac{ydy}{(4+y^2)} = \dfrac{xdx}{(1-x^2)}$

$\int \dfrac{ydy}{4+y^2} - \int \dfrac{xdx}{1-x^2} = 0$

$\dfrac{y\frac{dv}{2y}}{u} - \dfrac{x\frac{dv}{-2x}}{v} = 0$

$u = 4 + y^2$
$du = 2ydy$
$v = 1 - x^2$
$dv = -2xdx$

$\dfrac{1}{2}\int\dfrac{1}{u} + \dfrac{1}{2}\int\dfrac{1}{v} = 0$

$\left[\dfrac{1}{2}\ln|4+y^2| + \dfrac{1}{2}\ln|1-x^2| = c\right]2$

$\ln|4+y^2| + \ln|1-x^2| = 2c$

$e^{ln}|(4+y^2)(1-x^2)| = 2c$

$\boxed{(4+y^2)(1-x^2) = C_r}$

2. $y' - 2y = y^2$; $y = 3$ when $x = 0$

$\left[\dfrac{dy}{dx} - 2y = y^2\right]$

$\left[\dfrac{dy}{dx} = y^2 + 2y\right]dx$

$\dfrac{dy=(y^2+2y)dx}{y^2+2y}$

$\dfrac{dy}{y^2+2y} = dx$

$u = y + 1$
$du = 1$

$\int\dfrac{dy}{y^2+2y} - \int dx = 0$

$\int\left[\dfrac{1}{(u+1)(u-1)} = \dfrac{A}{(u+1)} + \dfrac{B}{(u-1)}\right](u+1)(u-1)$

$1 = A(u-1) + B(u+1)$

$u = 1$ $\qquad\qquad u = -1$

$\dfrac{1}{2} = b + \dfrac{B(2)}{2}$ $\qquad -\dfrac{1}{2} = \dfrac{A(-2)}{-2} = 0$

$\dfrac{1}{2} = B$ $\qquad\qquad -\dfrac{1}{2} = A$

$\int -\dfrac{1}{2(u+1)} + \int\dfrac{1}{2(u+1)} + x = 0$

$v = u \pm 1$
$dv = 1$

$+\dfrac{1}{2}\left[\int\dfrac{1}{u-1}du - \int\dfrac{1}{u+1}du\right] + x = 0$

$\dfrac{1}{2}\int\dfrac{1}{v} - \int\dfrac{1}{v} + x = 0$

$\left[\dfrac{1}{2}\,[ln|v| - ln|v|] + x = c\right]2$

$ln|u-1| - ln|u+1| + 2x = 2c$

$ln|y+1-1| - ln|y+1+1| + 2x = 2c$

$ln|y| - ln|y+2| + 2x = 2c$

$ln\left|\dfrac{y}{y+2}\right| + 2x = 2c$

$ln\left|\dfrac{3}{3+2}\right| + 2(0) = 2c$

$\dfrac{ln\left|\dfrac{3}{5}\right| + 2c}{2} = c$

$ln\left|\dfrac{y}{y+2}\right| - 2\left[\dfrac{ln\left|\frac{3}{5}\right|}{2}\right] = 2x$

$ln\left|\dfrac{y}{y+2}\right| - ln\left|\dfrac{3}{5}\right| = 2x$

$e^{ln}\left|\dfrac{5y}{(y+2)^3}\right| = e^{2x}$

$\left[\dfrac{5y}{(y+2)^3} = e^{2x}\right](y+2)3$

$\boxed{5y = 3(y+2)e^{2x}}$

3. $dr = b\,(\cos\theta\ dr + r\sin\theta\ d\theta)$ (b is constants)

$$dr = b\,(\cos\theta\ dr + \sin\theta\ d\theta)\ (b\ is\ constant)$$

$$\overline{}\, b$$

$\dfrac{dr}{b} - \dfrac{\cos\theta}{1} = r\sin\theta\ d\theta$

$\left[\dfrac{dr - b\cos\theta dr}{b} = r\sin\theta\ d\theta\right]b$

$dr - b\cos\theta dr = br\sin\theta\ d\theta$

$[dr\,(1 - b\cos\theta) = br\sin\theta\ d\theta]\dfrac{1}{r(1-b\cos\theta)}$

$\dfrac{dr}{r} = \dfrac{b\sin\theta d\theta}{1-b\cos\theta}$

$\int\dfrac{dr}{r} - \int\dfrac{b\sin\theta d\theta}{1-b\cos\theta} = 0 \qquad u = 1b\cos\theta$

$\qquad\qquad\qquad\qquad\qquad\qquad \dfrac{du}{b\sin\theta} = \dfrac{b\sin\theta d\theta}{1-b\cos\theta}$

$\ln|r| - \dfrac{b\sin\theta\cdot\frac{du}{b\sin\theta}}{u} = 0$

$\ln|r| - \ln|1 - b\cos\theta| = c$

$e^{\ln}\left|\dfrac{r}{1-b\cos\theta}\right| = e^c$

$\left[\dfrac{r}{1-b\cos\theta} = c_1\right]1 - b\cos\theta$

$\boxed{r = c_1(1 - b\cos\theta)}$

4. $y\,\ln x\,\ln y\ dx + dy = 0$

$$\overline{\dfrac{y\,\ln x\,\ln y\ dx = -dy}{y\ln y}}$$

$\ln x\ dx = -\dfrac{dy}{y\ln y} \qquad u = \ln x$

$\qquad\qquad\qquad\qquad\quad dv = dx$

$\int \ln x\ dx + \int\dfrac{dy}{y\ln y} = 0$

$\int u\,dv = uv - \int v\,du$

$\ln x\,(x) - \int y\dfrac{1}{x}\ dx$

$x\ln x - x + \int\dfrac{dy}{y\ln y} = 0 \qquad u = \ln y$

$\qquad\qquad\qquad\qquad\qquad\quad \dfrac{du}{1/y} = \dfrac{1/y\ dy}{1/y}$

$x\ln x - x + \dfrac{\frac{dy}{1/y}}{yu} = 0$

$x\ln x - x + \dfrac{1}{u} = 0$

$x\ln x - x + \ln|\ln y| = c$

$\boxed{x\ln x + \ln|\ln y| = c + x}$

5. $(y')^2 = \frac{1-y^2}{1-x^2}$; $y = \frac{1}{2}$ when $x = 1$.

$\left[\frac{dy^2}{dx^2} = \frac{(1-y^2)}{(1+x^2)}\right]\frac{dx^2}{(1-y^2)}$

$\sqrt{\frac{dy^2}{(1-y^2)}} = \sqrt{\frac{dx^2}{(1-x^2)}}$

$\frac{dy}{(1-y^2)} = \frac{dx}{(1-x^2)}$

$\int\frac{dy}{\sqrt{1-y^2}} - \int\frac{dx}{\sqrt{1-x^2}} = 0$

$\sin^{-1}(y) - \sin^{-1}(x) = c$

$\sin^{-1}\left(\frac{1}{2}\right) - \sin^{-1}(1) = c$

$\quad -\frac{1}{3}\Pi = c$

$\left[\sin^{-1}(y) - \sin^{-1}(x) = -\frac{1}{3}\Pi\right] - 1$

$-\sin^{-1}(y) + \sin^{-1}(x) = \frac{1}{3}\Pi$

$$\boxed{\sin^{-1}(x) + \sin^{-1}(y) = \frac{1}{3}\Pi}$$

6. $x\,e^y\,dy + \frac{x^2+1}{y}\,dx = 0$

$\left[x\,e^y\,dy = -\frac{x^2+1}{y}\,dx\right]\frac{y}{x}$

$y e^y\,dy = -\frac{x^2+1}{x}\,dx = 0$

$\int y e^y\,dy + \int\frac{x^2+1}{x}\,dx = 0 \qquad \int u\,dv = uv - \int v\,du$

$\searrow \int\frac{x^2}{x} + \frac{1}{x}$

$y e^y - \int e^y\,dy + \int x\,dx + \frac{dx}{x} = 0$

$y e^y - e^y + \frac{x^2}{x} + \ln x = c$

$$\boxed{e^y(y-1) + \frac{1}{2}x^2 + \ln x = c}$$

$\frac{u\,dv}{}$

7. $xy^3\,dx + (y+1)e^{-x}\,dy = 0$

$[xy^3\,dx = -(y+1)e^{-x}\,dy]y^{-3}e^x$

$x e^x\,dx = -(y+1)y^{-3}dy$

$\int x e^x\,dx + \int(y+1)y^{-3}dy = 0$

$x e^x - \int e^x dx + \int\frac{1}{y^2}dy + \frac{1}{y^3}dy = c$

$x e^x - e^x - \frac{1}{y} - \frac{1}{2y^2} = c$

$$\boxed{e^x(x-1) = \frac{1}{y} + \frac{1}{2y^2} + c}$$

8. $2xyy' = 1 + y^2$; $y = 3$ when $x = 2$.

$\left[2xy\dfrac{dy}{dx} = 1 + y^2\right]\dfrac{dx}{x}$

$\dfrac{2ydy = 1 + y^2\frac{dx}{x}}{1 + y^2}$

$\dfrac{2ydy}{1 + y^2} = \dfrac{dx}{x}$ $u = 1 + y^2$

                    $\dfrac{du}{2y} = dy$

$\int \dfrac{2ydy}{1+y^2} - \int \dfrac{dx}{x} = 0$

$\dfrac{2y\frac{du}{2y}}{u} - lnx = 0$

$ln|1 + y^2| - ln|x| = c$

$e^{ln\left|\frac{1+y^2}{x}\right|} = e^c$

$\dfrac{1+y^2}{x} = c$

$\dfrac{1+3^2}{2} = c$

$\dfrac{10}{2} = c$

$5 = c$

$\left[\dfrac{1+y^2}{x} = 5\right]x$

$1 + y^2 = 5x$

$\boxed{y^2 = 5x - 1}$

9. $(4z + x^2z)\,dz + (1 + z^2)dx - (yz^2 + y)(4 + x^2)dy = 0$

$\dfrac{(4 + x^2)dz + (1 + z^2)dx - y(z^2 + 1)(4 + x^2)dy = 0}{(4 + x^2)(1 + z^2)}$

$\dfrac{zdz}{(1+z^2)} + \dfrac{dx}{4+x^2} - ydy = 0$

$\int \dfrac{zdz}{(1+z^2)} + \int \dfrac{dx}{4+x^2} = \int ydy$ $u = 1 + z^2$

                                 $du = 2z$

$\int \dfrac{z\frac{du}{2z}}{u} + \int \dfrac{dx}{4 + x^2} = \dfrac{y^2}{2}$

$\left[\dfrac{1}{2}ln|1 + z^2| + \dfrac{1}{2}tan^{-1}\dfrac{x}{2} = \dfrac{y^2}{2} + c\right.$

$ln|1 + z^2| + tan^{-1}\dfrac{x}{2} = y^2 + 2c$

$\boxed{ln|1 + z^2| + tan^{-1}\dfrac{x}{2} - y^2 = c_2}$

10. $dz - (4x^3 - 4x^3 + x^3z^2)dx + (4y - 1)(2 - z)^2dy = 0$

$dz - x^3(4 - 4z + z^2)dx + (4y - 1)(z - 2)^2dy = 0$

$\dfrac{dz - x^3(z - 2)^2dx + (4y - 1)(z - 2)^2dy = 0}{(z - 2)^2}$

$\int \dfrac{dz}{(z-2)^2} - \int x^3dx + \int(4y - 1)dy = 0$ $u = z - 2$

                                           $du = 1$

$\dfrac{dzu^{-2+1}}{-2+1} - \dfrac{x^{3+1}}{3+1} + \dfrac{4y^{1+1}}{1+1} - \dfrac{y^{0+1}}{0+1} = 0$

$\dfrac{1}{u} - \dfrac{x^4}{4} + \dfrac{4y^2}{2} - y = 0$ $\longrightarrow$ $\boxed{\dfrac{1}{z-2} - \dfrac{1}{4}x^4 + 2y^2 - y = c}$

CHAPTER 3: HOMOGENOUS DIFFERENTIAL EQUATION

3.5 SUPPLEMENTARY PROBLEM

Find the complete or particular solution of the following differential equation:

1. $ydx + (2x + 3y)dy = 0$

$$x = vy \qquad v = \frac{x}{y}$$

$$dx = vdy + ydv$$

$$\frac{y(vdy + ydv)}{y} + \frac{(2vy + 3y)}{y}dy = \frac{0}{y}$$

$$vdy + ydv + 2vdy + 3dy$$

$$3vdy + 3dy + ydv = 0$$

$$3dy(v + 1) + ydv = 0$$

$$\int \frac{3dy}{y} + \int \frac{dv}{v+1} \qquad \begin{array}{l} let\ u = v + 1\ dv \\ du = dv \end{array}$$

$$3\int \frac{1}{y}dy + \int \frac{1dv}{u}$$

$$3 \ln|y| + \ln|v + 1| = c$$

$$\ln y^3 + \ln|v + 1| = c$$

$$\ln(y^3 \bullet v + 1) = c$$

$$e^{\ln}y^3\left(\frac{x+y}{y}\right) = e^c$$

$$y^2(x + y) = c$$

$$\boxed{xy^2 + y^3 = c}$$

2. $(2xy + y^2)dx - 2x^2\,dy = 0; y = e\ x = e$

$$y = vx \qquad v = \frac{x}{y}$$

$$dy = vdx + xdv$$

$$(2x(xv) + (xv)^2)dx - 2x^2(xdv + vdx) = 0$$

$$\frac{x^2(2v + v^2)dx}{x^2} - \frac{x^2(2x\,dv + 2v\,dx)}{x^2} = 0$$

$$2v\,dx + v^2\,dx - 2x\,dv - 2v\,dx = 0$$

$$\int \frac{dx}{2x} - \int \frac{dv}{v^2} = 0$$

$$\frac{1}{2}\int \frac{1}{u}dx - \int v^{-2+1}dv$$

$$2\left[\frac{1}{2}\ln|x| + \frac{1}{v} = c\right]2$$

$$\ln x + \frac{2}{v} = 2c$$

$$\ln x + \frac{2}{y/x} = 2c$$

$$\ln x + \frac{2x}{y} = 2c$$

$$\frac{\ln x + \frac{2x}{y} = 2c}{2}$$

$$c = \frac{\ln x + \frac{2x}{y}}{2} = \frac{3}{2}$$

$$\boxed{y\ln x - 2x = 3y}$$

3. $y(x^2 + xy - 2y^2)dx + x(3y^2 - xy - x^2)dy = 0$

$y = xv \qquad\qquad v = \dfrac{y}{x}$

$dy = xdv + vdx$

$xv(x^2 + x^2v - 2x^2v - 2x^2v^2)dx + x(3x^2v^2 - x^2v - x^2)dy = 0$

$\dfrac{x^3(v + v^2 - 2x^3dx) + x^2(3v^2 - v - 1)dy = 0}{x^3}$

$(v + v^2 - 2x^2)\,dx + (3v^2 - v - 1)\,(xdv + vdx)$

$vdx + v^2\,dx - 2x^3dx + (3v^2 - v - 1)(xdv) + 3v^3\,dx - v^2\,dx - vdx = 0$

$v^3\,dx + (3v^2 - v - 1)x\,dv = 0$

$v^3\,dx + x(3v^2 - v - 1)\,dv = 0$

$\int \dfrac{dx}{x} + \int \left[\dfrac{3v^3 - v - 1}{v^3}\right]dv = 0$

$\ln x + 3\ln v + \dfrac{1}{v} + \left(\dfrac{1}{2v^2}\right) = 0$

$\ln x + \ln\dfrac{y}{x} + \dfrac{x}{y} + \dfrac{x^2}{2y^2} = c$

$\ln x + \ln x \left(\dfrac{y}{x}\right)^3 + \dfrac{x}{y} + \dfrac{x^2}{2y^2} = c$

$\left[\ln\left|\dfrac{x}{y^3}\right| + \dfrac{2y^2x + x^2y}{2y^2} = c\right]2y^2$

$2y^2 \ln\left|\dfrac{y^3}{x^2}\right| + 2yx + x^2 = 2y^2c$

4. $\left(x + y\sin\dfrac{y}{x}\right)dx - x\sin\dfrac{y}{x}\,dy = 0$

$y = xv \qquad\qquad v = \dfrac{y}{x}$

$dy = xdv + vdx$

$(x + vx\sin v)\,dx - x\sin v\,(xdv + vdx) = 0$

$dx + v\sin v\,dx - x\sin v\,dv - v\sin v\,dx = 0$

$\dfrac{dx - x\sin v\,dv}{x} = 0$

$\int \dfrac{dx}{x} - \int \sin v\,dv = 0$

$\ln x - (-\cos v) = 0$

$\ln x + \cos v\, C$

$\boxed{\ln x + \cos\dfrac{v}{x} = c}$

5. $y^2 dy = x(x \, dy - y \, dx)e^{x/y}$

$y^2 dy = uy \, (uy \, dy - y \, (udy + ydu)) \, e^u$

$y^2 dy = y^2(u^2 dy - u^2 dy)e^u$

$dy = -uy \, due^u = 0$

$\dfrac{dy}{y} + ue^u = 0$

$\displaystyle\int \dfrac{dy}{y} + \int ue^u \, du$

$\ln y + ue^u - e^u = c$

$\dfrac{\ln y}{c} = -ue^u + e^u$

$\dfrac{\ln y}{c} = e^{x/y}\left(-\dfrac{x}{y} + 1\right)$

$\boxed{\dfrac{y \ln y}{c} = e^{x/y}(-x + y)}$

6. $xy \, dx + 2(x^2 + 2y^2)dy = 0; x = 0 \ y = 1$

$(uy) y \, (vdy + ydu) + 2 \, ((uy)^2 + 2y^2)dy = 0$

$y^2 u \, (udy + ydu) + y^2 \, 2 \, (u^2 + 2)dy = 0$

$u^2 \, dy + uydu + (2u^2 + 4)dy = 0$

$\displaystyle\int \dfrac{dy}{y} + \int \dfrac{udu}{(3u^2 + 4)} = 0$ $\text{let } u = 3u^2 + 4$

$\ln y + \dfrac{1}{6}\displaystyle\int \dfrac{du}{u} = c$ $du = 6u$

$e^{ln}y^6 \cdot 4 = e^c$

$y^6\left(3\dfrac{x^2}{y^2} + 4\right)$

$(1)^6 \left(3\dfrac{(0)^2}{(1)^2} + 4\right) = c = 4$ $\longrightarrow$ $\boxed{y^4(3x^2 + 4y^2) = 4}$

7. $y' = \dfrac{y}{x} - \cot\dfrac{y}{x}$

$\dfrac{dy}{dx} = \dfrac{y}{x} - \cot\dfrac{y}{x} = 0$

$(vdx + xdv) = (v \cot v) \, dx = 0$

$vdx + xdv = vdx - \cot v \, dx$

$xdv = -\cot v \, dx = 0$

$\displaystyle\int \dfrac{dv}{\cot v} = -\int \dfrac{dx}{x}$ $\text{let } \cot v = \dfrac{\cos v}{\sin v}$

$\displaystyle\int \dfrac{\sin v}{\cos}$ $\text{let } u = \cos v \ \ du = -\sin v$

$-\dfrac{du}{u} = -\displaystyle\int \dfrac{dx}{x} = 0$

$-\ln u = -\ln x - \ln c$

$e^{-\ln u} = e^{-\ln x \bullet c}$

$-u = -x \bullet c$

$x \bullet c = \cos\dfrac{y}{x}$

$\boxed{\cos\dfrac{y}{x} = Cx}$

8. $xy' = y - \sqrt{x^2 - y^2}$

$x\dfrac{dy}{dx} = y - \sqrt{x^2 - y^2}$

$x(vdx + xdv) = (vx) - \sqrt{x^2 - v^2x^2}\,dx$

$vdx + xdv = vdx - \sqrt{1 - v^2}\,dx$

$xdv = -\sqrt{1 - v^2}$

$\dfrac{dv}{\sqrt{1-v^2}} + \dfrac{dx}{x}$

$\displaystyle\int \dfrac{dv}{\sqrt{1-v^2}} + \int \dfrac{dx}{x} = 0$

$\sin^{-1}\dfrac{v}{1} + \ln x = c$

$v = \sin(c - \ln x)$

$\dfrac{y}{x} = \sin(c - \ln x)$

$\boxed{y = x\sin(c - \ln x)}$

9. $x\,dy - y\,dx = \sqrt{x^2 - y^2}\,dx;\ \ x = \dfrac{1}{2}\ y = 0$

$x(vdx + xdv - (vx)dx = \sqrt{x^2 - v^2x^2}\,dx$

$x\,[(vdx + vdx - vdx] = x\sqrt{1 - v^2}\,dx$

$vdx + xdv - vdx = \sqrt{1 - v^2}\,dx$

$\dfrac{dv}{\sqrt{1 - v^2}} - \dfrac{dx}{x}$

$\displaystyle\int \dfrac{dv}{\sqrt{1-v^2}} - \int \dfrac{dx}{x} = C$

$\ln\left|v + \sqrt{1 - v^2}\right| - \ln x = \ln C$

$e^{\ln\left|\frac{v+\sqrt{1-v^2}}{x}\right|} = e^{\ln c}$

$\dfrac{y}{x} + \dfrac{\sqrt{x^2-y^2}}{x^2}{x}$

$\dfrac{\left(y + \dfrac{\sqrt{x^2 - y^2}}{x^2}\right)}{x^2}$

$\dfrac{(0 + \sqrt{1/2^2 + 0})}{1/2^2} = C$

$\dfrac{\frac{1}{2}}{\frac{1}{4}}\Big|_{\frac{1}{4}}^{\frac{1}{2}} = 2 = C$

$y + \sqrt{x^2 - y^2} = 2x^2$

$\left[\sqrt{x^2 - y^2} = 2x^2 - y\right]^2$

$x^2 + y^2 = 4x^4 - 4x^2 + y^2$

$0 = 4x^2 - 4y - 1$

$\boxed{-4x^2 + 4y + 1 = 0}$

10. $y^2 + x^2y' = xyy'$

$y^2 + x^2\dfrac{dy}{dx} = xy\dfrac{dy}{dx}$

$y^2 = xy\dfrac{dy}{dx} - x^2\dfrac{dy}{dx}$

$y^2 = (xy - x^2)\dfrac{dy}{dx}$

$y^2(udy + ydu) = ((uy) - (uy)^2)\,dy$

$y^2(udy + ydu) = y^2(u - u^2)\,dy$

$ydu = -u^2dy$

$\displaystyle\int \dfrac{du}{u^2} + \int \dfrac{dy}{y} = C$

$-\dfrac{1}{u} + \ln y = C$

$-\dfrac{x}{y} + \ln y = C$

$\ln y = \ln^C \cdot \dfrac{y}{x}$

$e^{\ln y} = e^{\ln c} \cdot e^{\frac{y}{x}}$

$\boxed{y = Ce^{\frac{y}{x}}}$

CHAPTER 4

EXACT DIFFERENTIAL EQUATIONS

4.3 SUPPLEMENTARY PROBLEMS

In the following differential equations, test for exactness and determine the solution of each:

1. $(6x + y^2)\, dx + y(2x - 3y)\, dy = 0$

$M = (6x + y^2)$; $\quad \dfrac{\partial M}{\partial y} = 0 + 2y$

$N = y(2x - 3y)$; $\quad \dfrac{\partial N}{\partial x} = 2y - 0$

The equation is **exact.**

$F_m = \int (6x\,dx + y^2\,dx) = 3x^2 + xy^2 + Qy$

$F_n = \int (2xy - 3y^2)\, dy = xy^2 - y^3 + Rx$

$Qy = -y^3$, $\qquad Rx = 3x^2$

Answer: $\boxed{3x^2 + xy^2 - y^3 = C}$

2. $(2xy - \tan y)\, dx + (x^2 - x\sec^2 y)\, dy = 0$

$M = (2xy - \tan y)$; $\quad \dfrac{\partial M}{\partial y} = 2x - \sec^2 y$

$N = (x^2 - x\sec^2 y)$; $\quad \dfrac{\partial N}{\partial x} = 2x - \sec^2 y$

The equation is **exact.**

$F_m = \int (2xy\,dx - \tan y\,dx) = x^2 y - x \tan y + Qy$

$F_n = \int (x^2 - x\sec^2 y)\, dy = x^2 y - x \tan y + Rx$

$Qy = 0$, $\qquad Rx = 0$

Answer: $\boxed{x^2 y - x \tan y = C}$

3. $\left(e^x + \ln y + \dfrac{y}{x}\right) dx + \left(\dfrac{x}{y} + \ln x + \sin y\right) dy = 0$

$M = \left(e^x + \ln y + \dfrac{y}{x}\right)$; $\quad \dfrac{\partial M}{\partial y} = 0 + \dfrac{1}{y} + \dfrac{1}{x}$

$N = \left(\dfrac{x}{y} + \ln x + \sin y\right)$; $\quad \dfrac{\partial N}{\partial x} = \dfrac{1}{y} + \dfrac{1}{x} + 0$

The equation is **exact.**

$F_m = \int \left(e^x dx + \ln y\,dx + \dfrac{y}{x}\,dx\right) = e^x + x\ln y + y\ln x + Qy$

$F_n = \int \left(\dfrac{x}{y} + \ln x + \sin y\right) dy = x\ln y + y\ln x - \cos y + Rx$

$Qy = -\cos y$, $\qquad Rx = e^x$

Answer: $\boxed{e^x + x\ln y + y\ln x - \cos y = 0}$

4. $(x + \sqrt{y^2 + 1})\, dx - \{y\frac{xy}{\sqrt{y^2+1}}\}\, dy = 0$

$M = (x + \sqrt{y^2 + 1})$; $\frac{\partial M}{\partial y} = 0 + \frac{y}{\sqrt{y^2 + 1}}$

$N = y\frac{xy}{\sqrt{y^2+1}}$; $\frac{\partial N}{\partial x} = 0 + \frac{y}{\sqrt{y^2 + 1}}$

The equation is **exact.**

$F_m = \int (x + \sqrt{y^2 + 1})\, dx = \frac{x^2}{2} + x\sqrt{y^2 + 1} + Qy$

$F_n = \int -(y\frac{xy}{\sqrt{y^2+1}})\, dy = -\frac{y^2}{2} + x\int \frac{y}{\sqrt{y^2+1}}\, dy$

Let $u = y^2 + 1$, $du = 2y$, $= \frac{1}{2} \cdot \frac{u^{-\frac{1}{2}}}{-\frac{1}{2}+1}$, $= \frac{1}{2} \cdot \frac{2}{1} u^{\frac{1}{2}}$, $= u^{\frac{1}{2}}$

$= -\frac{y^2}{2} + x\sqrt{y^2 + 1} + Rx$

$Qy = -\frac{y^2}{2}$, $Rx = \frac{x^2}{2}$

$= 2[\frac{x^2}{2} + x\sqrt{y^2 + 1} - \frac{y^2}{2}] = C$

Answer: $x^2 + 2x\sqrt{y^2 + 1} - y^2 = C$

5. $3y (x^2 - 1)\, dx + (x^3 + 8y - 3x)\, dy = 0$; $y = 1$ when $x = 0$

$M = (3yx^2 - 3y)$; $\frac{\partial M}{\partial y} = 3x^2 - 3$

$N = (x^3 + 8y - 3x)$; $\frac{\partial N}{\partial x} = 3x^2 + 0 - 3$

The equation is **exact.**

$F_m = \int (3yx^2 - 3y)\, dx = yx^3 - 3xy +$

$F_n = \int (x^3 + 8y - 3x)\, dy = yx^3 + 4y^2 - 3yx + Rx$

$Qy = 4y^2$, $Rx = 0$

$= yx^3 - 3xy + 4y^2 = C$; $y = 1$ $x = 0$

$= (1)(0)^3 - 3(0)(1) + 4(1)^2 + C = 4$, $= yx^3 - 3xy + 4y^2 = 4$

Answer: $xy(x^2 - 3) = 4(1 - y^2)$

6. $\frac{dy}{dx} - \frac{xdy}{y^2} = 0$

$M = \frac{dy}{dx}$; $\frac{\partial M}{\partial y} = -\frac{1}{y^2}$

$N = \frac{xdy}{y^2}$; $\frac{\partial N}{\partial x} = -\frac{1}{y^2}$

The equation is **exact.**

$F_m = \int \frac{dy}{dx} = \frac{x}{y}$

$F_n = -\int \frac{xdy}{y^2} = -\frac{1}{y^2} \cdot xy \;,\; = -\frac{x}{y} + Rx$

$Qy = 0$, $Rx = 0$

$= \frac{x}{y} + 0 = C$

Answer: $x = Cy$

---

7. $(x+3)^{-1} \cos y \, dx - \{\sin y \ln(5x + 15) - y^{-1}\} dy = 0$

$= \frac{\cos y \, dx}{x+3} - \{\sin y \ln(5x + 15) - \frac{1}{y}\} dy = 0$

$M = \frac{\cos y}{x+3}$; $\frac{\partial M}{\partial y} = -\frac{\sin y}{x+3}$

$N = -\{\sin y \ln(5x + 15) - y^{-1}\}$; $\frac{\partial N}{\partial x} = 0 - \sin y\left(\frac{1}{x+3}\right) + 0$

The equation is **exact.**

$F_m = \int \frac{\cos y \, dx}{x+3} = \cos y \ln (x+3) + Qy$

$F_n = -\int \{\sin y \ln(5x + 15) - \frac{1}{y}\} dy = \cos y \ln (5) + \cos y \ln (x+3) + \ln y + Rx$

$Qy = \cos y \ln (5) + \ln (y)$, $Rx = 0$

$= \cos y \ln (x+3) + \cos y \ln (5) + \ln (y) = C$

Answer: **$\cos y \ln (5x + 15) + \ln y = C$**

---

8. $\frac{x \, dy - y \, dx}{y^2} = x^3 dx$

$= \frac{x}{y^2} dy = (x^3 + \frac{y}{y^2}) dx$

$= -\frac{x}{y^2} dy + (x^2 + \frac{1}{y}) dx$

$N = -\frac{x}{y^2}$; $\frac{\partial N}{\partial x} = -\frac{1}{y^2}$

$M = x^2 + \frac{1}{y}$; $\frac{\partial M}{\partial y} = 0 - \frac{1}{y^2}$

The equation is **exact.**

$$F_n = -\int \left(\frac{x}{y^2}\right) dy = \frac{x}{y} + Rx$$

$$F_m = \int \left(x^3 + \frac{1}{y}\right) dx = \frac{x^4}{4} + \frac{x}{y} + Qy$$

$$Qy = 0 \qquad , \qquad Rx = \frac{x^4}{4}$$

$$= \frac{x^4}{4} + \frac{x}{y} = C$$

Answer: $x^4 + 4x = Cy$

5.8 SUPPLEMENTARY PROBLEMS

1. $(2x^2y - 2y^2 + 2xy)\,dx + (x^2 - 2y)\,dy = 0$

$M = 2x^2y - 2y^2 + 2xy$ $\qquad ; \quad \frac{\partial M}{\partial y} = 2x^2 - 4y + 2x$

$N = x^2 - 2y$ $\qquad\qquad ; \quad \frac{\partial N}{\partial x} = 2x - 0$

The equation is **non-exact.**

$F_x = \frac{1}{x^2} - 2y\,(2x^2 - 4y + 2x - 2x)$

$= \frac{2x^2 - 4y}{x^2 - 2y}$

$= \frac{2(x^2 - 2y)}{x^2 - 2y}$

$F_x = 2$

$\emptyset = e^{\int 2\,dx}$

$\emptyset = e^{2x}$

$= (2x^2ye^{2x} - 2y^2e^{2x} + 2xye^{2x})\,dx + (x^2e^{2x} - 2ye^{2x})\,dy = 0$

$\frac{\partial M}{\partial y} = (2x^2e^{2x} - 4ye^{2x} + 2xe^{2x})$

$\frac{\partial N}{\partial x} = x^2(2e^{2x}) + e^{2x}(2x) - 2y\,(2)\,e^{2x} = 2x^2e^{2x} + 2xe^{2x} - 4ye^{2x}$

EXACT

$F_m = \int (2y^2e^{2x} - 2y^2e^{2x} - 2yxe^{2x})\,dx = 2y\left(x^2\frac{e^{2x}}{2} - \frac{e^{2x}}{4}(2x)\right)$ $\qquad\qquad =$

$x^2\frac{e^{2x}}{2} - \frac{xe^{2x}}{2} = 2y\left(\frac{x^2e^{2x}}{2} - \frac{xe^{2x}}{2} + \frac{e^{2x}(1)}{4}\right) - 2y^2\frac{e^{2x}}{2} + 2y\left(\frac{xe^{2x}}{2} - \frac{e^{2x}(1)}{4}\right)$ $\qquad =$

$x^2e^{2x}y - xe^{2x}y + \frac{e^{2xy}}{2} - e^{2x}y^2 + xe^{2x}y - \frac{ye^{2x}}{2}$

$\qquad = x^2e^{2x}y - e^{2x}y^2 + Fy$

$F_n = \int (x^2e^{2x} - e^{2x}\,2y)\,dy = x^2e^{2x}y - e^{2x}y^2 + Gx$

$Fy = 0 \qquad , \qquad Gx = 0$

$F = x^2e^{2x}y - e^{2x}y^2 = C \qquad , \qquad = e^{2x}(x^2y - y^2) = C$

Answer: $\boxed{y(x^2 - y) = C\,e^{-2x}}$

2. $(x^2 + y^2)dx - xy\,dy = 0$

$M = x^2 + y^2$; $\quad \frac{\partial M}{\partial y} = 0 + 2y$

$N = -xy$ $\quad$; $\quad \frac{\partial N}{\partial x} = -y$

The equation is **non-exact.**

$F_x = \frac{1}{-xy}(2y + y) = \frac{3y}{-xy} = \frac{3}{-x}$

$\emptyset = e^{\int -\frac{3}{x}dx}$ $,= e^{-3\ln x}$ $\qquad ,= e^{\ln x^{-3}}$ $\qquad\qquad ,= x^{-3}$

$= (x^2 \cdot x^{-3} + y^2 x^{-3})dx - xy \cdot x^{-3}\,dy = 0$

$\frac{\partial M}{\partial y} = 0 + 2yx^{-3} = \frac{2y}{x^3}$

$\frac{\partial N}{\partial x} = -\frac{y}{x^2} = \frac{x^2(0) - y2x}{x^4} = \frac{2y}{x^3}$

EXACT

$F_m = \int \left(\frac{1}{x} + \frac{y^2}{x^3}\right)dx$

$\quad = \ln(x) + y^2 \int \frac{1}{x^3}dx \quad = \ln(x) + y^2\left(-\frac{1}{2x^2}\right)$

$\quad = \ln(x) - \frac{y^2}{2x^2} + Fy$

$F_n = \int \left(-\frac{y}{x^2}\right)dy = -\frac{y^2}{2x^2} + Gx$

$Fy = 0$ $\quad , \quad$ $Gx = \ln(x)$

Answer: $\ln(x) - \dfrac{y^2}{2x^2} = C$

---

3. $(x + 4y^3)dy - y\,dx = 0$

$M = x + 4y^3$; $\quad \frac{\partial M}{\partial y} = 1 + 0 = 1$

$N = -y$ $\quad$; $\quad \frac{\partial N}{\partial x} = -1$

$\quad = x\,dy + 4y^3\,dy - y\,dx = 0$

$\quad = -\frac{1}{y^2}[x\,dy - y\,dx + 4y^3\,dy] = 0$

$\quad = -\frac{x\,dy}{y^2} + \frac{y\,dx}{y^2} - \frac{4y^3}{y^2}dy = 0$

$\quad = -x\,dy + \frac{y\,dx}{y^2} - 4y\,dy = 0$

$\quad = d\left(\frac{x}{y}\right) - 4\int y\,dy = 0$

$\quad = \frac{x}{y} - 4\frac{y^2}{2} = C$

$$= \frac{x}{y} - 2y^2 = C$$

$$= \frac{x}{y} = C + 2y^2$$

Answer: $x = y(C + 2y^2)$

4. $(4xy + y^2)dx - 2(x^2 - y)\, dy = 0$

$M = 4xy + y^2$; $\frac{\partial M}{\partial y} = 4x + 2y$

$N = 2(x^2 - y)$; $\frac{\partial N}{\partial x} = -2x + 2y = -4x + 0$

The equation is **non-exact.**

$4xy\, dx + y^2 dy - 2x^2 dy + 2y\, dy = 0$

$4xy\, dx - 2x^2\, dy + y^2\, dx + 2y\, dy = 0$

$\frac{1}{2y^2}[2(2xy\, dx + x^2 dy) + y^2 dx + 2y\, dy = 0]$

$\frac{2(2xy\, dx - x^2 dy)}{2y^2} + \frac{y^2 dx}{2y^2} + \frac{2y\, dy}{2y^2} = 0$

$\frac{2xy\, dx - x^2 dy}{y^2} + \frac{dx}{2} + \frac{dy}{y} = 0$

$d\frac{x^2}{y} + \frac{dx}{2} + \frac{dy}{y} = 0$

$\frac{x^2}{y} + \frac{x}{2} + \ln(y) = C$

$2y\,[\frac{x^2}{y} + \frac{x}{y} + \ln(y)] = C$

Answer: $2x^2 + xy + 2y \ln(y) = Cy$

5. $(2x^2 y^{-2} + 1)dx + xy^{-1}dy = 0$

$\left(\frac{2x^2}{y^2} + 1\right)dx + \frac{x}{y}dy = 0$

$y^2\left[\left(\frac{2x^2}{y^2} + 1\right)dx + \frac{1}{y}\, dy = 0\right]$

$(2x^2 + y^2)dx + xy\, dy = 0$

$M = 2x^2 + y^2$; $\frac{\partial M}{\partial y} = 0 + 2y$

$N = xy$; $\frac{\partial N}{\partial x} = y$

The equation is **non-exact.**

$$F_x = \frac{1}{xy}(2y - y)$$

$$= \frac{y}{xy}$$

$$= \frac{1}{x}$$

$$\emptyset = e^{\int x^{-1}dx}$$

$$= e^{\ln x}$$

$$\emptyset = x$$

$$= (2x^2 \cdot x + xy^2)dx + x \cdot xy\, dy = 0$$

$$= (2x^3 + xy^2)\, dx + x^2 y\, dy = 0$$

$$M = 2x^3 + xy^2 \quad ; \quad \frac{\partial M}{\partial y} = 0 + 2xy$$

$$N = x^2 y \quad\quad ; \quad \frac{\partial N}{\partial x} = 2xy$$

EXACT

$$F_m = \int(2x^3 + xy^2)dx = \frac{2x^4}{4} + \frac{x^2}{2}y^2 + Fy$$

$$F_n = \int(x^2 y)dy = \frac{x^2 y^2}{2} + Gx$$

$$Fy = 0 \quad , \quad Gx = \frac{x^4}{2}$$

$$F = \frac{x^4}{2} + \frac{x^2 y^2}{2} + 0 = C$$

Answer: $x^2(x^2 y^2) = C$

6. $(y - xy^3)\, dx + x\, dy = 0$

$$M = y - xy^3 \quad ; \quad \frac{\partial M}{\partial y} = 1 - 3xy^2$$

$$N = x \quad\quad ; \quad \frac{\partial N}{\partial x} = 1$$

The equation is **non-exact.**

$$= \frac{(y - xy^3)}{y}n - \frac{xm}{x} = 1 - (1 - 3xy^2)$$

$$= (1 - xy^2)n - m = 3xy^2$$

$$= n - nxy^2 - m = 3xy^2$$

$$3xy^2 = -nxy^2 \quad\quad ; \quad n - m = 0$$

$$n = -3 \quad\quad\quad\quad ; \quad n = m = -3$$

$$\emptyset = x^m y^n$$

$$\emptyset = x^{-3}y^{-3}$$

$$\emptyset = (xy)^{-3}$$

$$\frac{x-xy^3}{(xy)^3}dx + \frac{x}{(xy)^3}dy = 0$$

$$\left(\frac{1}{x^3y^2} - \frac{1}{x^2}\right)dx + \frac{1}{x^2y^3}dy = 0$$

$$M = \frac{1}{x^3y^2} - \frac{1}{x^2} \quad ; \quad \frac{\partial M}{\partial y} = \frac{1}{x^3}\left(\frac{-2}{y^2}\right) - 0 = \frac{-2}{x^3y^3}$$

$$N = \frac{1}{x^2y^3} \quad ; \quad \frac{\partial N}{\partial x} = \frac{1}{y^3}\left(\frac{-2}{y^3}\right) = \frac{-2}{y^3x^3}$$

EXACT

$$F_m = \int \frac{1}{x^3y^2} - \frac{1}{x^2} \, dx = \frac{1}{y^2}\left(\frac{-1}{2x^2}\right) - \frac{-1}{x} + Fy$$

$$= \frac{-1}{2x^2y^2} + \frac{1}{x} + Fy$$

$$F_n = \int \frac{1}{x^2y^3} = \frac{1}{x^2}\left(\frac{-1}{2y^2}\right) + Gx$$

$$= \frac{-1}{2x^2y^2} + Gx$$

$$Fy = 0 \quad , \quad Gx = \frac{1}{x}$$

$$F = \left[\frac{-1}{2x^2y^2} + \frac{1}{x}\right] - 2 = C$$

$$\boxed{\text{Answer} = (xy)^{-2} = 2x^{-1} + C}$$

7. $y(y^2 + 1)dx + x(y^2 - 1)dy = 0$

$$y^3dx + ydx + xy^2dy - xdy = 0$$

$$y^3dx + xy^2dx + ydx - xdy = 0$$

$$y^{-2}[y^2(ydx + xdy) + ydx - xdy = 0]$$

$$(ydx + xdy) + \frac{ydx}{y^2} - \frac{xdy}{y^2} = 0$$

$$(ydx + xdy) + \frac{ydx - xdy}{y^2} = 0$$

$$\int d\,(xy) + \int d\left(\frac{x}{y}\right) = 0$$

$$\left[xy + \frac{x}{y} = C\right]y$$

$$xy^2 + \frac{x}{y} = Cy$$

$$\boxed{\text{Answer: } x(y^2 + 1) = Cy}$$

8. $(x^3 + xy^2 - y)\,dx + (y^3 + x^2y + x)\,dy = 0$

$x^3\,dx + xy^2\,dx - y\,dx + y^3\,dy + x^2y\,dy + x\,dy = 0$

$(x\,dy - y\,dx) + x^3\,dx + xy^2\,dx + y^3\,dy + x^2y\,dy = 0$

$(x\,dy - y\,dx) + x(x^2 + y^2)\,dx + y(y^2 + x^2)\,dy = 0$

$[(x\,dy - y\,dx) + x(x^2 + y^2)\,dx + y(y^2 + x^2)\,dy = 0](y^2 + x^2)^{-1}$

$x\,dy - \frac{y\,dx}{y^2x^2} + x\,dx + y\,dy = 0$

$\int d\left(\text{Arctan}\frac{y}{x}\right) + \int x\,dx + \int y\,dy = 0$

$\left[\text{Arctan}\frac{y}{x} + \frac{x^2}{2} + \frac{y^2}{2} = C\right]2$

$2\,\text{Arctan}\frac{y}{x} + x^2 + y^2 = 2C \quad \text{Let } 2C = C$

Answer: $\boxed{2\,\textbf{Arctan}\frac{y}{x} = C - x^2 - y^2}$

9. $2xy\,dx + (y^2 - x^2)\,dy = 0; y = 2, x = 0$

$M = 2xy \qquad ; \qquad \frac{\partial M}{\partial y} = 2x$

$N = y^2 - x^2 \quad ; \qquad \frac{\partial N}{\partial x} = 0 - 2x$

The equation is **non-exact.**

$Fy = \frac{1}{2xy}(-2x - 2x)$

$\quad = \frac{-4x}{2xy}$

$Fy = \frac{-2y}{y}$

$\emptyset = e^{\int Fy\,dy}$

$\quad = e^{-2\int \frac{dy}{y}}$

$\quad = e^{-2\ln y}$

$\emptyset = y^{-2}$

$\frac{2xy\,dx}{y^2} + \left(\frac{y^2}{y^2} - \frac{x^2}{y^2}\right)dy = 0$

$\frac{2x}{y}\,dx + \left(1 - \frac{x^2}{y^2}\right)dy = 0$

$M = \frac{2x}{y} \qquad ; \qquad \frac{\partial M}{\partial y} = \frac{-2x}{y^2}$

$N = 1 - \frac{x^2}{y^2} \; ; \qquad \frac{\partial N}{\partial x} = 0 + \frac{-2x}{y^2}$

EXACT

$$F_m = \int \frac{2x}{y}\, dx = \frac{x^2}{y} + Fy$$

$$F_n = \int 1 - \frac{x^2}{y^2} = y + \frac{x^2}{y} + Gx$$

$$Fy = y \quad , \quad Gx = 0$$

$$= y + \frac{x^2}{y} = C$$

$$= (2) + \frac{0}{2} = C = 2$$

$$F = y\left[y + \frac{x^2}{y} = 2\right]$$

Answer: $y^2 + x^2 = 2y$

6.4. SUPPLEMENTARY PROBLEM

1. $x y' + y = x^4 - 3x$

   $[\, x y' + y = x^4 - 3x \,]\, x^{-1}$

   $y' + \dfrac{y}{x} = \dfrac{x^4 + 3x}{x}$

   $\dfrac{dy}{dx} + \dfrac{y}{x} = x^3 - 3$

   $Qx = x^3 - 3$

   $Px = \dfrac{1}{x}$

   $I.F = e^{\int pxdx}$

   $\quad = e^{\int \frac{dx}{x}}$

   $\quad = e^{lnx}$

   $\quad = x$

   $yx = \int x(x^3 - 3)dx + C$

   $yx = \int (x^4 - 3x)dx + C$

   $yx = \dfrac{x^5}{5} - \dfrac{3x^2}{2} + C$

   $[\, yx = \dfrac{x^5}{5} - \dfrac{3x^2}{2} + C \,]\dfrac{1}{x}$

   $\boxed{y = \dfrac{x^4}{5} - \dfrac{3x}{2} + Cx^{-1}}$

2. $x^2y' - 2xy = x^4 + 3$; y=2 when x=1

$[\, x^2y' - 2xy = x^4 + 3 \,]\, x^{-2}$

$y' - \dfrac{2xy}{x^2} = \dfrac{x^4+3}{x^2}$

$y' - \dfrac{2y}{x} = \dfrac{x^4+3}{x^2}$

$Px = -\dfrac{2}{x}$

$Qx = \dfrac{x^4+3}{x^2}$

$I.F = e^{-2\int \frac{dx}{x}}$

$= e^{-2lnx}$

$= x^{-2}$

$\dfrac{y}{x^2} = \int \dfrac{x^4+3}{x^2 \cdot x^2}\, dx + C$

$\dfrac{y}{x^2} = \int \left(1 + \dfrac{3}{x^4}\right) dx + C$

$\dfrac{y}{x^2} = x + 3\left(\dfrac{-1}{3x^3}\right) + C$

$\dfrac{y}{x^2} = x - \dfrac{1}{x^3} + C$

Value of C;

$\dfrac{2}{1} = 1 - \dfrac{1}{1} + C$

$C = 2$

$[\, \dfrac{y}{x^2} = x - \dfrac{1}{x^3} + C \,]\, x^2$

$y = x^3 - x^{-1} + Cx^2$

$\boxed{y = x^3 + 2x^2 - x^{-1}}$

3. $(\sin 2\theta - 2p \cos \theta) \, d\theta = 2dp$

$[\, (\sin 2\theta - 2p \cos \theta) \, d\theta = 2dp \,] \dfrac{1}{d\theta}$

$\sin 2\theta - 2p \cos \theta = 2\dfrac{dp}{d\theta}$

$\sin 2\theta = 2\dfrac{dp}{d\theta} + 2p \cos \theta$

$\dfrac{dp}{d\theta} + p\cos\theta = \dfrac{\sin 2\theta}{2}$

$P\theta = \cos\theta$

$Q\theta = \dfrac{\sin 2\theta}{2}$

$I.F = e^{\int \cos\theta \, d\theta}$

$\quad = e^{\sin\theta}$

$Pe^{\sin\theta} = \displaystyle\int e^{\sin\theta} \cdot \dfrac{\sin 2\theta}{2} \, d\theta + C$

$Pe^{\sin\theta} = \displaystyle\int e^{\sin\theta} \cdot \dfrac{2\sin\theta\cos\theta}{2} \, d\theta + C$

$Pe^{\sin\theta} = \displaystyle\int e^{\sin\theta} \cdot \sin\theta\cos\theta \, d\theta + C$

$Let \ u = \sin\theta$

$du = \cos\theta \, d\theta$

$Pe^{u} = \displaystyle\int e^{u} \cdot u \, du + C$

$Pe^{u} = ue^{u} - \displaystyle\int e^{u} du + C$

$Pe^{u} = ue^{u} - e^{u} + C$

$[\, Pe^{u} = ue^{u} - e^{u} + C \,] \dfrac{1}{e^{u}}$

$P = u - 1 + Ce^{-u}$

$\boxed{P = \sin\theta - 1 + Ce^{-\sin\theta}}$

4. $y \, dx - 4x \, dy = y^6 \, dy$; $x = 4$ when $y = 1$

$y \, dx = (y^6 + 4x) \, dy$

$[\, y \, dx = (y^6 + 4x) \, dy] \, \dfrac{1}{dx} \cdot \dfrac{1}{y^6 + 4x}$

$[\dfrac{y}{y^6 + 4x} = \dfrac{dy}{dx}]^{-1}$

$\dfrac{y^6 + 4x}{y} = \dfrac{dx}{dy}$

$\dfrac{dx}{dy} - \dfrac{4x}{y} = y^5$

$Py = -\dfrac{4}{y}$

$Qy = y^5$

$I.F = e^{-4 \int \frac{dx}{dy}}$

$\quad = e^{-4lny}$

$\quad = y^{-4}$

$\dfrac{x}{y^4} = \displaystyle\int \dfrac{y^5}{y^4} \, dy + C$

$\dfrac{x}{y^4} = \dfrac{y^2}{2} + C$

Value of C;

$\dfrac{4}{1} = \dfrac{1}{2} + C$

$C = \dfrac{7}{2}$

$[\dfrac{x}{y^4} = \dfrac{y^2}{2} + \dfrac{7}{2}] \, 2y^4$

$\boxed{2x = y^4(y^2 + 7)}$

5. $t \frac{dx}{dt} = 6t\, e^{2t} + x(2t - 1)$

y=x

x=t

$\left[t \frac{dx}{dt} = 6t\, e^{2t} + x(2t - 1) \right] \frac{1}{t}$

$\frac{dx}{dt} = 6e^{2t} + x(\frac{2t - 1}{t})$

$\frac{dx}{dt} = x\left(\frac{2t - 1}{t}\right) = Ce^{2t}$

$Pt = -\frac{2t - 1}{t}$

$Qt = Ce^{2t}$

$I.F = e^{\int Ptdt}$

$= e^{-\int(2 - \frac{1}{2})dt}$

$= e^{-(2t - Int)}$

$= e^{-2t} \cdot e^{Int}$

$= \frac{e^{Int}}{e^{2t}}$

$= \frac{t}{e^{2t}}$

$\frac{xt}{e^{2t}} = \int \frac{t}{e^{2t}} \cdot 6e^{2t}\, dt + C$

$\frac{xt}{e^{2t}} = 6 \int t\, dt + C$

$\frac{xt}{e^{2t}} = 3\, t^2 + C$

$\left[\frac{xt}{e^{2t}} = 3\, t^2 + C \right] e^{2t}$

$\boxed{xt = (3t^2 + C)\, e^{2t}}$

6. $(y - x + xy \cot x) \, dx + x \, dy = 0$

$$[(y - x + xy \cot x) \, dx + x \, dy = 0] \, \frac{1}{dx} \cdot \frac{1}{x}$$

$$\frac{(y - x + xy \cot x)}{x} + \frac{dy}{dx} = 0$$

$$\frac{y}{x} + y \cot x + \frac{dy}{x} = 0 + 1$$

$$y\left(\frac{1}{x} + \cot x\right) + \frac{dy}{dx} = 1$$

$$\frac{dy}{dx} + y\left(\frac{1}{x} + \cot x\right) = 1$$

$$Px = \frac{1}{x} + \cot x$$

$$Qx = 1$$

$$I.F = e^{\int \frac{1}{x} + \cot x \, dx}$$

$$= e^{\ln x + \ln \sin x}$$

$$= e^{\ln(x \cdot \sin x)}$$

$$= x \sin x$$

$$yx \sin x = \int x \sin x \, dx + C$$

$$yx \sin x = -x \cos x - \int -\cos dx + C$$

$$yx \sin x = -x \cos x + \sin x + C$$

$$\boxed{xy \sin x = C + \sin x - x \cos x}$$

7. $v\,dx + (2x + 1 - vx)dv = 0$

$$[v\,dx + (2x + 1 - vx)dv = 0]\frac{1}{dv}$$

$$\frac{dx}{dv} + \frac{2x}{v} + \frac{1}{v} - x = 0$$

$$\frac{dx}{dv} + \frac{2x}{v} - x = 0 - \frac{1}{v}$$

$$\frac{dx}{dv} + x\left(\frac{2}{v} - 1\right) = -\frac{1}{v}$$

$$Pv = \frac{2}{v} - 1$$

$$Qv = -\frac{1}{v}$$

$$I.F = e^{\int \frac{2}{v} - 1\,dv}$$

$$= e^{2lnv - v}$$

$$= e^{lnv^2} \cdot e^{-v}$$

$$= \frac{v^2}{e^v}$$

$$\frac{xv^2}{e^v} = \int \frac{v^2}{e^v} \cdot \frac{-1}{v}\,dv + C$$

$$\frac{xv^2}{e^v} = -\int \frac{v}{e^v}\,dv + C$$

$$\frac{xv^2}{e^v} = (-v \cdot -e^{-v}) - \int -e^{-v} - dv + C$$

$$\frac{xv^2}{e^v} = ve^{-v} - (-e^{-v}) + C$$

$$\frac{xv^2}{e^v} = \frac{v}{e^v} + \frac{1}{e^v} + C$$

$$\left[\frac{xv^2}{e^v} = \frac{v}{e^v} + \frac{1}{e^v} + C\right]e^v$$

$$\boxed{xv^2 = v + 1 + Ce^v}$$

8. $y' - \frac{2y}{x} = x^2 e^x$; $x = 1$ when $y = 0$

$Px = -\frac{2}{x}$

$Qx = x^2 e^x$

$I.F = e^{-2 \int \frac{dx}{x}}$

$\quad = e^{-2lnx}$

$\quad = x^{-2}$

$\frac{y}{x^2} = \int x^{-2} \cdot x^2 e^x \, dx + C$

$\frac{y}{x^2} = e^x + C$

$\frac{0}{1} = e^1 + C$

$C = -e$

$\frac{y}{x^2} = e^x - e$

$\left[\frac{y}{x^2} = e^x - e \right] x^2$

$\boxed{y = x^2(e^x - e)}$

9. $\dfrac{dy}{dt} + \dfrac{s}{t} = \cos t + \dfrac{\sin t}{t}$

$Pt = \dfrac{1}{t}$

$Qt = \cos t + \dfrac{\sin t}{t}$

$I.F = e^{\int \frac{dt}{t}}$

$\quad = e^{\ln t}$

$\quad = t$

$st = \displaystyle\int t \left(\cos t + \dfrac{\sin t}{t}\right) dt + C$

$st = \displaystyle\int (t\cos t + \sin t) dt + C$

$st = t\sin t - \displaystyle\int \sin t \, dt + \displaystyle\int \sin t \, dt + C$

$st = t \sin t + C$

$[\, st = t \sin t + C \,]\dfrac{1}{t}$

$\boxed{s = \sin t + \dfrac{C}{t}}$

10. $nx \dfrac{dy}{dx} + 2y = xy^{n+1}$ (n is constant)

$$\left[nx \dfrac{dy}{dx} + 2y = xy^{n+1} \right] \dfrac{1}{nx}$$

$$\dfrac{dy}{dx} + \dfrac{2y}{nx} = \dfrac{y^{n+1}}{n}$$

$$Px = \dfrac{2}{nx}$$

$$Qx = \dfrac{1}{n}$$

$$N = n + 1$$

$$Prx = -\dfrac{2}{x}$$

$$Qrx = -1$$

$$I.F = e^{\int -2\frac{dx}{x}}$$

$$= e^{-2lnx}$$

$$= x^{-2}$$

$$z = y^{-n}$$

$$\dfrac{z}{x^2} = \int \dfrac{-1}{x^2} dx + C$$

$$\dfrac{z}{x^2} = \dfrac{1}{x} + C$$

$$[\dfrac{1}{y^n x^2} = \dfrac{1}{x} + C\,]\, y^n x^2$$

$$1 = xy^n + Cy^n x^2$$

$$\boxed{Cx^2 y^n + xy^n - 1 = 0}$$

7.4. SUPPLEMENTARY PROBLEM

1. $2x\,dy - y(x+1)dx + 6y^3\,dx = 0$

$$[\,2xdy - y(x+1) + 6y^3dx = 0\,]\frac{1}{dx}$$

$$\left[\frac{2xdy}{dx} - y(x+1) + 6y^3 = 0\right]\frac{1}{2x}$$

$$\frac{dy}{dx} - y\left[\frac{x}{2x} - \frac{1}{2x}\right] + \frac{6y^3}{2x} = 0$$

$$\left[\frac{dy}{dx} - y\left(\frac{1}{2} + \frac{1}{2x}\right) = -\frac{3y^3}{x}\right]y$$

Let $v = y^{-2}$

$dv = -2y^{-3}dy$

$$\frac{dv}{-2} = y^{-3}dy$$

$$y^{-3}\frac{dy}{dx} - v\left(\frac{1}{2} - \frac{1}{2x}\right) = -\frac{3}{x}$$

$$\left[\frac{dv}{-2dx} - v\left(\frac{1}{2} - \frac{1}{2x}\right) = -\frac{3}{x}\right] - 2$$

$$\frac{dv}{dx} - y^2\left(-1 - \frac{1}{x}\right) = \frac{6}{x}$$

$$Px = 1 + \frac{1}{x}$$

$$Qx = \frac{6}{x}$$

$$I.F = e^{\int 1 + \frac{1}{x}dx}$$

$I. F = e^{x+lnx}$

$I. F = xe^x$

$vxe^x = \int \frac{6}{x}(xe^x)dx + C$

$vxe^x = \int 6e^{-x} dx + C$

$vxe^x = 6(e^x) + C$

$[\,vxe^x = 6e^x + C\,]e^{-x}$

$vx = 6 + Ce^{-x}$
$[\,y^{-2}x = 6 + Ce^{-x}\,]y^2$

$$\boxed{x = y^2\,[\,6 + Ce^{-x}\,]}$$

2. $2x^3y' = y(y^2 + 3x^2)$

$\left[\,2x^3\dfrac{dy}{dx} = y^3 + 3x^2y\,\right]\dfrac{1}{2x^3}$

$\dfrac{dy}{dx} = \dfrac{y^3}{2x^3} + \dfrac{3y}{2x}$

$[\dfrac{dy}{dx} + \left(-\dfrac{3y}{2x}\right) = \dfrac{y^3}{2x^3}]y^{-3}$

$y^{-3}\dfrac{dy}{dx} + \left(-\dfrac{3y}{2x}\right)y^{-2} = \dfrac{1}{2x^3}$

Let $v = y^{-2}$
$dv = -2y^{-3}\,dy$
$-\dfrac{dv}{2} = y^{-3}dy$

$\left[\dfrac{dv}{2dx} + \left(-\dfrac{3}{2x}\right)v = \dfrac{1}{2x^3}\right] - 2$

$$\frac{dv}{dx} + \frac{3}{x}v = -\frac{1}{x^3}$$

$$Px = \frac{3}{x}$$

$$Qx = -\frac{1}{x^3}$$

$$I.F = e^{3\int \frac{1}{x}dx}$$
$$= e^{3lnx}$$
$$= x^3$$

$$vx^3 = \int -\frac{1}{x^3}(x^3)dx + C$$

$$vx^3 = \int -1\ dx + C$$

$$vx^3 = -x + C$$

$$[y^{-2}x^3 = -x + C\ ]y^2$$

$$\boxed{x^3 = y^2(C - x)}$$

3. $dy + y\ dx = 2xy^2\ e^x\ dx$

$$[\ dy + y\ dx = 2xy^2\ e^x\ dx\ ]\frac{1}{dx}$$

$$\left[\frac{dy}{dx} + y = 2xy^2\ e^x\right]\frac{1}{y^2}$$

$$y^{-2}\frac{dy}{dx} + y^{-1} = 2xe^x$$

Let $v = y^{-1}$
$dv = -y^{-2}dy$
$-dv = y^{-2}$

$$-\left[-\frac{dv}{dx} + v = 2xe^x\right] -$$

$$\frac{dv}{dx} - v = -2xe^x$$

$$Px = -1$$
$$Qx = -2xe^x$$
$$I.F = e^{-\int 1\,dx}$$
$$= e^{-x}$$

$$ve^{-x} = \int -2xe^x(e^{-x})dx + C$$
$$ve^{-x} = \int -\frac{2x^2}{2} + C$$
$$ve^{-x} = -x^2 + C$$
$$y^{-1}e^{-x} = x^2 + C$$
$$[\frac{1}{ye^x} = -x^2 + C\,]ye^x$$

$$\boxed{1 = ye^x[-x^2 + C]}$$

4. $$dx + 2x\frac{dy}{y} = 2x^2y^2\,dy$$

$$[\,dx + 2x\frac{dy}{y} = 2x^2y^2\,dy\,]\frac{1}{dy}$$

$$[\frac{dx}{dx} + \frac{2x}{y} = 2x^2y^2\,]x^{-2}$$

$$x^{-2}\frac{dx}{dy} + \left(\frac{2}{y}\right)x^{-1} = 2y^2$$

Let $v = x^{-1}$
$$dv = -x^{-2}\,dx$$
$$-dv = x^{-2}\,dx$$

$$-[\frac{-dv}{dy} + \left(\frac{2}{y}\right)v = 2y^2$$

$$\frac{dv}{dy} - \left(\frac{2}{y}\right)v = -2y^2$$

$$Px = -\frac{2}{y}$$

$$Qx = -2y^2$$

$$I.F = e^{-\int \frac{2}{y}dy}$$

$$= e^{-2lny}$$

$$= y^{-2}$$

$$vy^{-2} = \int -2y^2(y^{-2})dy + C$$

$$vy^{-2} = \int -2dy + C$$

$$vy^{-2} = -2y + C$$

$$\boxed{x^{-1}y^{-2} = -2y + C}$$

5. $2\,wt\,\dfrac{dt}{dw} = t^2 - 2w^3$

$$\left[2\,wt\,\frac{dt}{dw} = t^2 - 2w^3 \right]\frac{1}{2\,wt}$$

$$\frac{dt}{dw} = \frac{t}{2w} - \frac{w^2}{t}$$

$$[\,\frac{dt}{dw} = \frac{t}{2w} - \frac{w^2}{t}\,]\,t$$

$$t\,\frac{dt}{dw} - \frac{t^2}{2w} = -w^2$$

Let $v = t^2\,dt$

$dv = 2t\,dt$

$$\frac{dv}{2} = t\,dt$$

$$\left[\frac{dv}{2dw} - \frac{v}{2w} = -w^2\right]2$$

$$\frac{dv}{dw} - \frac{v}{w} = -2w^2$$

$$Px = -\frac{1}{w}$$

$$Qx = -2w^2$$

$$I.F = e^{-\int \frac{1}{w}dw}$$

$$= e^{-lnw}$$

$$= w^{-1}$$

$$v(w^{-1}) = \int -2w^2(w^{-1})dw + C$$

$$v(w^{-1}) = \int -2w\ dw + C$$

$$v(w^{-1}) = -w^2 + C$$

$$[\,t^2 w^{-1} = -w^2 + C\,]w$$

$$\boxed{t^2 = w\,(C - w^2)}$$

CHAPTER 9: Special Second-Ordered Equation

SUPPLEMENTARY PROBLEMS

1. $\frac{d}{dx}\left(\frac{dy}{dx}\right) = 6x+3$

$\int \frac{d}{dx}\left(\frac{dy}{dx}\right) = \int 6x + 3$

$v = 3x^2 + 3x + c_1$

$\frac{d}{dx} = 3x^2 + 3x + c_1$

$\boxed{y = x^3 + 3x^2/2 + c_1\,x + c_2}$

2. $\frac{d}{dx}\{x + \frac{dy}{dx} + (1 + x)y\} = 12$; y=0, d''/dx=0 when x=1 (particular solution)

$\int \frac{d}{dx}\left[x\frac{dy}{dx} + (1 + x)\right] = 12x + c$

$x\frac{d}{dx} + (1 + x)y = 12x + c$

x=0 ; y=0 ; $\frac{d}{dx} = 0$

0 + 2(0) = 12(1) +c_1

c_1 = 12

$x\frac{d}{dx} + (1 + x)y = 12x - 12$

$x\frac{d}{dx} + \left(\frac{1}{x} + x\right)y = 12 - \frac{12}{x}$

$\theta = e^{\int \frac{1}{x}+1}$

$\theta = e^{\,lnx+x}$

$\theta = xe^x$

$yxe^x = \int \left(12 - \frac{12}{x}\right)(xe^x)dx + c_2$

$yxe^x = \int (12xe^x - 12e^x)dx + c_2$

$yxe^x = 12xe^x - 12e^x - 12e^x + c_2$

u = x ; du = e^x

D = dx ; v = e^x

CHAPTER 9: Special Second-Ordered Equation

$yxe^x = 12xe^x - 12e^x\text{-}12e^x + c_2$

$yx = 12xe^x - 24e^x + c_2e^{-x}$

$yx = 12x - 24 + c_2e^{-x}$

$0 = 12 - 24 + c_2e^{-1}$

$c_2 = 12e$

$$\boxed{yx = 12x - 24 + 12e^{1-x}}$$

3. –

4. $y'' + 2y' - 6x - 3 = 0$

$\dfrac{dp}{dx} + 2p = 6x + 3$

P=2

$Q = 6x + 3$

$\theta = e^{\int 2dx}$

$\quad = e^{2x}$

$\int (6x + 3)e^x dx + c = pe^{2x}$

$pe^{2x} = 6\left(\dfrac{xe^{2x}}{2} - \dfrac{e^{2x}}{4}\right) + 3xe^x + c$

$pe^{2x} = 3xe^{2x} - \dfrac{3e^{2x}}{2} + 3xe^{2x} + c$

$p = 3x + \dfrac{3}{2} + c_1e^{-2x}$

$\dfrac{dy}{dx} = 3x + \dfrac{3}{2} + c_1e^{-2x}$

$$\boxed{y = \dfrac{3x^2}{2} + \dfrac{3x}{2} + c_1e^{-2x} + c_2}$$

CHAPTER 9: Special Second-Ordered Equation

5. $yy'' + 2(y')^2 = 0$

$yp\dfrac{dp}{dy} + 2p^2$

$y\dfrac{dp}{dy} + 2p = 0$

$\dfrac{dp}{dy} + \dfrac{2p}{y} = 0$

$lny + 2lnp = c$

$yp^2 = c$

$p = \dfrac{c}{y^{\frac{1}{2}}}$

$\dfrac{dy}{dx} = \dfrac{c}{y^{\frac{1}{2}}}$

$dyy^{\frac{1}{2}} = cdx$

$\dfrac{y^{\frac{1}{2}+1}}{\frac{1}{2}} + 1 = c_1x + c_2$

$y^{\frac{3}{2}} = cx$

$$\boxed{y^3 = c_1x + c_2}$$

6. $yy'' - (y')^2 + y^1 = 0$

$yp\dfrac{dp}{dy} - p^2 + p = 0$

$\dfrac{dp}{dy} - \dfrac{p}{y} = -\dfrac{1}{y}$

$p = -\dfrac{1}{y}$

$\theta = e^{\int -\frac{1}{y}dy} = e^{lnxy^{-1}} = y^{-1}$

$Q = -\dfrac{1}{y}$

$py^{-1} = -\int \dfrac{1}{y}(y^{-1})dy + c$

$py^{-1} = \dfrac{1}{y} + c$

$p = \dfrac{1}{y^2} + \dfrac{c}{y}$

CHAPTER 9: Special Second-Ordered Equation

$$\left(cy - y + \frac{1}{cy+1}\right)dy = dx$$

$$\frac{c_1}{2}y^2 - y + 1\ln y + \frac{1}{c_1} = x$$

$$\frac{(c_1)^2}{2}y^2 - c_1 y + \ln y + c_1 = x + c$$

1. **For a certain curve, the point of contact of each tangent to it bisects the part of the tangent terminating on the coordinate axes. Find the equation of the curve.**

Solution:

$$\frac{dy}{dx} = \frac{\frac{y}{2} - y}{\frac{x}{2} - 0}$$

$$\frac{dy}{dx} = \frac{-\frac{y}{2}}{\frac{x}{2}}$$

$$\frac{dy}{dy} = \frac{-dx}{x}$$

$$\ln y = -\ln x + C$$

$$\ln x \, y = C$$

$xy = C$

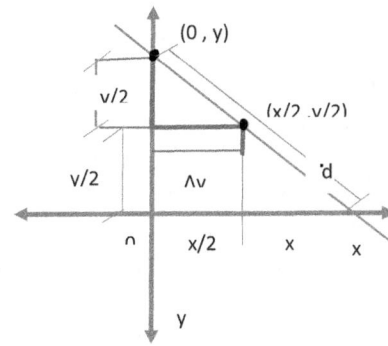

2.   **Find the equation of the curve so drawn that every point on it is equidistant from the orig**
and the intersection of the X-axis with the normal to the curve at the point.

Solution:

$\frac{dy}{dx} = \frac{-1}{\left(\frac{y-0}{x-x}\right)}$

$x^2 + y^2 = (x - x_1)^2 + y^2$

$D = D$

$x^2 = (x - x_1)^2$

$\frac{dy}{dx} = \frac{-(x-x_1)}{y}$

$y\,dy = -dx(x)$

$y\,dy = -dx(x)$

$\frac{y^2}{2} + \frac{x^2}{2} = C$

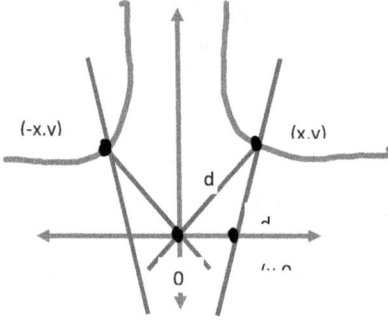

(-x.v)   (x.v)

d

0

3. **The area bounded by a curve, the X-axis, a fixed coordinate, and a variable ordinate is proportional to the difference between the ordinates. Find the equation of the curve.**

Solution:

$$A = ydx$$

$$A = ydx = k(y_1 - y) \ \ Let: \ dy = (y_1 - y)$$

$$ydx = kdy$$

$$dx = k\frac{dy}{y}$$

$$\frac{x + C}{k} = \ln y$$

$$y = e^{\frac{x+C}{k}}$$

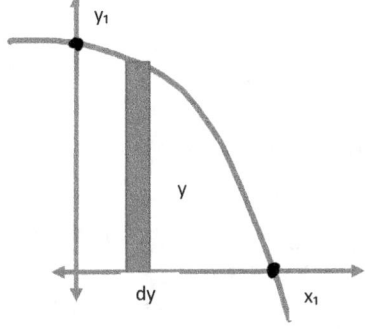

4. *Find the orthogonal trajectories of the ellipses 4x2 + y2= ex.*

Solution:

$4x^2 + 2y^2 = Cx$

$8x = 2yy' = \frac{4x}{2} = \frac{y^2}{2x}$

$2x + yy' = \frac{y^2}{2x}$

$y' = \frac{y^2}{2xy} - \frac{2x}{y}$

$y' = \frac{y}{2x} - \frac{2x}{y}$

$2y' = \frac{y^2 - 4x^2}{2xy}$

$-\frac{1}{y_1} = y^1 r$

$y' = \frac{2xy}{4x^2 - y^2}$

$v = \frac{x}{y}$

$\frac{y^2}{2\frac{x^2}{y^2} - 1} = C$

$C = \frac{4x^2 + 2y^2}{x}$

$(4x^2 - y^2)dy = 2xy\,dx$

Let: $x = uy$

$dx = udy + ydv$

$(4u^2 - 1)dy = 2v^2 dy + 2vydv$

$(2u^2 - 1)dy = 2vydv$

$\frac{dy}{y} = \frac{2u}{2u^2 - 1}dv$

$\ln y = \frac{2}{4}\ln|2u^2 - 1| + C$

$\frac{y^2}{2u^2 - 1} = C$

$$\frac{y^4}{2x^2 - y^2} = C$$

5. *Determine the isogonal trajectories of the circles x2 + y2= c if the angles of intersecti are to be 45 degrees.*

Solution:

$x^2 + y^2 = C$

$\int (x + y)dy + f(x)$

$1 = \frac{y'r - yy'}{1 + y'ry'y}$

$2x + 2yy' = 0$

$y' = -\frac{x}{y}$

$\tan(45) = 1$

$= xy + y^{\frac{2}{2}} + f(x)$

$F = \int (y - x)dx + g(y)$

$= yx - \frac{x^2}{2} + g(y)$

$F =$

$$1 = \frac{y'r + \frac{x}{y}}{1 + y'r\left(-\frac{x}{y}\right)}$$

$$1 + y'r\left(-\frac{x}{y}\right) = y'r + \frac{x}{y}$$

$$y'r\left(1 + \frac{x}{y}\right) = 1 - \frac{x}{y}$$

$$f(x) = -x^{\frac{2}{2}}$$

$$g(y) = y^{\frac{2}{2}}$$

$$yx - \frac{x^2}{2} + \frac{y^2}{2} = C$$

$$\frac{dy}{dx}\left[\frac{x+y}{y}\right] = \frac{y-x}{y}$$

$$dy(x + y) = (y - x)dx$$

$$(x + y)dy + (y - x)dx = 0$$

$$\frac{\partial M}{\partial x} = 1 \qquad \frac{\delta N}{\delta y} = 1 \text{``}$$

6.   **A weight of 450 N falls from rest. If the resistance of the air is proportional to the speed, and if the limiting speed is 52.5 m/s, find the speed after 5 sec.**

Solution:

\\\\/

$t = \varsigma$

KV

$$450 - kv = \frac{450}{g}\frac{dv}{dt}$$

$$450 - k(52.5) = 0$$

$$k = \frac{450}{52.5}$$

$$k = \frac{60}{7}$$

$$\left(450 - \frac{60v}{7}\right) = \frac{gdt}{450dt}$$

$$\frac{dv}{450 - kv} = \frac{gdt}{450}$$

$$\frac{\ln|450 - kv|}{-k} = \frac{gdt}{450} + C$$

$$\frac{\ln|450|}{-k} = \frac{g(0)}{450} + C$$

$$C = -\frac{\ln|450|}{k}$$

$$\frac{\ln|450 - kv|}{-k} = \frac{gt}{450} - \frac{\ln|450|}{k}$$

$$\ln|450 - kv| = -\frac{kgt}{450} + \ln|450|$$

$$-\frac{kv}{450} = e - \frac{kgt}{450} - 1$$

$$v = \left[1 - e - \frac{kgt}{450}\right]450$$

$$k = \frac{60}{7} \quad @\, t = 5 \quad V = 31.87\,\frac{m}{s}$$

5

7. **A 45 N weight is projected downward with an Initial speed of 4.6 m/s. If the air resistance is proportional to the speed and if the limiting speed is double the initial speed, how far has body traveled when it has reached a speed of 6 m/s?**

Solution:

$$w - kv = \frac{w}{g}\left(\frac{dv}{dt}\right)$$

$$@ \frac{dv}{dt} = 0, \quad k = 2(4.6)\backslash$$

$$w - k(9.2) = 0$$

$$k = \frac{w}{9.2}$$

$$w - \frac{wv}{9.2} = \frac{w}{g}\frac{dv}{dt}$$

$$1 - \frac{v}{9.2} = \frac{dv}{gdt}$$

$$\frac{dv}{1-\frac{v}{9.2}} = gdt$$

$$-9.2 \ln\left|1 - \frac{v}{9.2}\right| = gt + C$$

0, $\quad v = 4.6$

$$-9.2 \ln\left|1 - \frac{1}{21}\right| = C$$

$$C = 6.38$$

$$S = \left[\frac{0 + e^{\frac{gt}{9.2}-0.69}}{\frac{g}{9.2}}\right] 9.2$$

$$S_1 = 4.32m \quad \text{(Initial Position)}$$

$$-9.2 \ln\left|1 - \frac{v}{9.2}\right| = gt + 6.38$$

$$@ v = 6, \quad t =?$$

$$t = 0.34$$

$$-9.2 \ln\left|1 - \frac{v}{9.2}\right| = gt + 6.38$$

$$1 - \frac{v}{9.2} = e^{-\frac{gt}{9.2}-0.69}$$

$$v = \left[1 - e^{-\frac{gt}{9.2}-0.69}\right] 9.2$$

$$\int \frac{ds}{dt} = \int \left[1 - e^{-\frac{gt}{9.2}-0.69}\right] 9.2$$

$$S = \frac{t + e^{-\frac{gt}{9.2}-0.69}}{\frac{g}{9.2}} \qquad @ t =$$

$$S = \left[\frac{t + e^{-\frac{gt}{9.2}-0.69}}{\frac{g}{9.2}}\right] 9.2$$

$$S_2 = 6.14 \, m \quad \text{(Final Position)}$$

$$\Delta S = 6.14 - 4.32$$

$$\Delta S = 1.81 \, m$$

8. **A body falls from rest against a resistance proportional to the cube of the speed at a instant. If the limiting speed is 3 m/s. find the time required to attain a speed of 2 m/s.**

Solution:

$$1 - \frac{v^3}{3^3} = \frac{w}{g}\frac{dw}{dt}$$

$$\frac{g}{3^3} dt = \frac{dv}{3^3 - v^3} \qquad = \frac{gt}{3^3} \qquad = \left(\frac{1}{27(3-v)} + \frac{\frac{1}{27}v + \frac{2}{9}}{9+3v+v^2}\right) dv$$

$$\frac{A}{3-v} + \frac{Bv+D}{A+3v+v^2} = \frac{1}{3^3 - v^3} \qquad Let; \quad v = 9 + 3v + v^2 \quad dv = 3 + 2v$$

6

$$9A + 3D = 1$$
$$3A + 3B - D = 0$$
$$-B + A = 0$$
$$B = A$$

$$6A - D = 0$$
$$D = 6A$$

$$9A + 18A = 1$$
$$A = \frac{1}{27} = B$$
$$D = \frac{2}{9}$$

$$\frac{gt}{3^3} = -\frac{1}{27}\ln|3 - v| + \int \frac{\frac{1}{27}v + \frac{2}{9}}{9+3v+V^2}$$
$$3 = \frac{2}{9}A$$
$$A = \frac{2}{27}$$
$$\frac{gt}{3^3} = -\frac{1}{27}1 - |3 - v| + \frac{1}{54}\int \frac{2v+12}{9+3v+v^2}$$
$$\frac{1}{54}\int \frac{2x+3}{9+3v+v^2} + \frac{9}{9+3v+v^2}$$
$$\frac{1}{54}\ln \quad |9 + 3v + v^2| + \frac{9}{v^2+3v+\frac{9}{4}+\frac{27}{4}}$$
$$\frac{9}{\frac{3\sqrt{3}}{2}}\arctan\left|\frac{v+\frac{3}{2}}{\frac{3\sqrt{3}}{2}}\right| + C\int \frac{9}{\left(v+\frac{3}{2}\right)^2+\frac{27}{4}}$$
$$\frac{gt}{3^3} = -\ln|3 - v| + \frac{1}{54}\ln|9 + 3v + v^2| + \frac{6}{\sqrt{3}(54)}\left|\frac{2v+3}{3\sqrt{3}}\right| + C$$
$$@\, t = 0,\ v = 0 \qquad C = -0.011$$
$$@\, v = 2,\ t =? \qquad t = 0.28\text{ sec}$$

9. **A body above the surface of the earth is pulled toward the center of the earth with a force proportional to the reciprocal of the square of the distance of the body from the center. If the radius of the earth is 6375 km, find the velocity of the body as it strikes the surface of the earth if it falls from rest at a distance of four times the earth's radius measured from the center of the earth. What is the velocity if it falls from an infinite distance?**

Solution:

$$\frac{K}{D^2} = \frac{W}{g}\frac{dv}{dt}$$

$$-\frac{R^2}{D} = \frac{1}{g}\frac{v^2}{2} + \frac{R}{4}$$
$$@\, D = R$$

$$@\, D = R \qquad \frac{dv}{dt} = -g$$

$$\frac{K}{D^2} = \frac{W}{g}(-g)$$
$$V = \sqrt{\left(\frac{3R}{4}\right)2g}$$

$$K = -R^2(W)$$
$$V = 9.7\ \frac{m}{sec}$$

$$-\frac{R^2}{D^2} = \frac{1}{g}\frac{dv}{dt}\frac{dD}{dD}$$
$$\frac{R^2}{D} = \frac{1v^2}{2g} + C$$

$$-\frac{R^2}{D^2} = \frac{1}{g}v\frac{dv}{dD}$$
$$-\frac{R^2}{D} = \frac{1}{g}\frac{v^2}{2} + C$$
$$\frac{R^2}{D} =$$

$$\frac{1}{2g}v^2 \qquad \sqrt{R(2)(g)} = V$$

$$@\, v = 0 \qquad D = 4$$
$$V = 11.18\ \frac{km}{sec}$$

$$-\frac{R^2}{4R} = C = \frac{1}{4R}$$

10. **A boat with its load weighs 1430 N. If the force exerted upon the boat by the motor in the direction of motion is equivalent to a constant force of 67 N, and if the resistance to motion is equal numerically to twice the speed, find the speed after 10 sec. Assume the boat starts from rest.**

7

Solution:

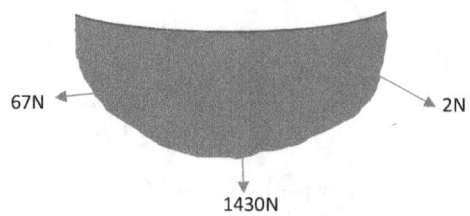

67N

2N

1430N

$$\frac{1430}{g}\frac{dv}{dt} = 67 - 2v$$

$$\frac{dv}{67 - 2v} = \frac{gdt}{1430}$$

$$-\frac{1}{2}\ln|67 - 2v| = \frac{gt}{1430} + C$$

$$C + 67 - 2v = e^{-\frac{2gt}{1430}} + C$$

@ $t = 0$ $v = 0$ $c = 67$

$$\frac{67 - Ce^{-\frac{2gt}{1430}}}{2} = V$$

$$\frac{67 - 67e^{-\frac{2gt}{1430}}}{2} = V$$

$$V = 4.29\ \frac{m}{s}$$

11.   *A certain radioactive material follows the law of exponential change and has a half-life of 38 hours. Find how long it takes for 90% of the radioactivity to be dissipated.*

Solution:

$$\frac{ds}{dt} = ks$$

$$\frac{ds}{s} = kdt$$

$$\ln s = kt + C$$

$$s = Ce^{kt}$$

$$\frac{p_0}{2} = p_0 e^{k(38)}$$

$$0.1\ p_0 = p_0 e^{kt}$$

$$0.1 = e^{kt}$$

$$1(0.1) = kt$$

$$s = e^{kt+C}$$

$$\frac{\ln(0.1)(38)}{\ln\left(\frac{1}{2}\right)} = t$$

$$T = 126\ tons$$

8

$$\frac{1}{2} = p_o e^{k(38)}$$

$$\ln \left|\frac{1}{2}\right| = 38k$$

$$K = \frac{\ln \dfrac{1}{2}}{38}$$

12. **A bacterial population follows the law of exponential growth. If between noon and 2 p.m. the population triples, at what time should the population become 100 times what it was at noon? At 10 a.m. what percentage was present?**

Solution:

$$3p_0 = pe^{k2}$$

$$\ln|3| = 2k$$

$$k = \frac{\ln|3|}{2}$$

$$100p_0 = p_o e^{kt}$$

$$\ln|100| = kt$$

$$t = \frac{\ln|100|}{\ln|3|} \quad (2)$$

$$t = 8.22 \; pm$$

13. **A thermometer reading 75°F is taken out where the temperature is 20° F. The reading is 30°F 4 minutes later, find (a) the thermometer reading 7 minutes after the thermometer was brought outside, and (b) the time taken for the reading to drop from 75°F to within a half degree of the air temperature or 20.5°F.**

Solution:

$$\frac{d_{tb}}{dt} = k(tb - -tm)$$

$tm = temp. \; of \; the \; medium$
$tb = temp. \; anytime$
$K = constant$

A.) $Tb - 20 = 55e^{kt}$

$$Tb - 20 = 55e^{\ln\left|\frac{30-20}{55}\right| \frac{(7)}{4}} = 22.7$$

$$\frac{d_{tb}}{dt} = k(tb - tm)$$

B.) $205 - 20 = 55e^{\ln\left|\frac{30-20}{55}\right|\frac{(T)}{4}}$

$$\frac{d_{tb}}{tb-tm} = kdt$$

$$T = \frac{\ln\left|\frac{205-20}{55}\right|}{4} \div \frac{\ln|30-20|}{55} \quad (T)$$

$$tb - tm = ce^{kt}$$

$$T = 11.03 \; min$$

$$75 - 20 = ce^{k(0)}$$

$$30 - 20 = ce^{k(4)}$$

9

$$30 - 20 = 55e^{4k}$$

$$\frac{\ln\left|\dfrac{30-20}{55}\right|}{4} = k = -0.43$$

14. At a certain time, a thermometer reading 70°F is taken outdoors where the temperature 15°F. Five minutes later, the thermometer reading is 45° F. After another 5 minutes, t thermometer is brought back indoors where the temperature is fixed at 70° F. What is t thermometer reading ten minutes after the thermometer is brought back indoors? When will t reading, to the nearest degree, return back to its original reading of 70°F?

Solution:

$70 - 15 = ce^{kt}$

$55 = C$ @ $t = 0$

$45 - 15 = 55e^{kt}$

$k = -0.121$

$Th = 15 = 55e^{k(10)}$

$Th = 31.4$

B.) $\dfrac{dtb}{dt} = -k(70 - T)$

$-\ln|70 - T| = -kt + C$

$-\ln|70 - 31.4| = -0.21 + C$

$-3.65 = C$

$-\ln|70 - T| = -0.21 + (-3.65)$

$T = 70$

A.) $31.4 - 70 = ce^{k(10)}$

$C = -38.6$

$T - 70 = -38.6e^{kt}$

$T = 70 - 38.6e^{k(10)}$

$T = 58.5\ ^{\circ}F$

15. A tank contains 200 liters of fresh water. Brine containing 2.5 N/liter of dissolved salt ru into the tank at the rate of 8 liters/min and the mixture kept uniform by stir- ring runs out at liters/min. Find the amount of salt when the tank contains 240 liters of brine. The concentration salt in the tank after 25 minutes amounts to how much?

Solution:

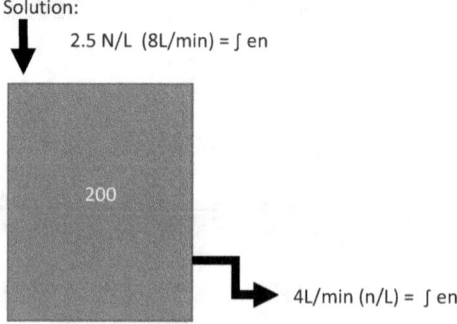

2.5 N/L (8L/min) = ∫ en

200

4L/min (n/L) = ∫ en

$$C = \frac{S}{200 + (8-4)(t)}$$

$$C = \frac{S}{200 + (4t)}$$

$$\frac{ds}{st} = \frac{20N}{min} - \frac{S}{200+4}\left(\frac{n}{L}\right)(4)\frac{L}{min}$$

$$\frac{ds}{st} = 20 - \frac{4S}{200+4t}$$

$$\frac{ds}{st} + \frac{4S}{200+4t} = 20$$

$$\theta = e^{\int \frac{4S}{200+4t}dt}$$

$$\theta = e^{\ln|200+4t+1|}$$

$$\theta = 200 + 4t$$

$$\int (200+4t) = 20\int (200+4t)dt + C$$

$$\int (200+4t) = 20(200+2t^2) + C$$

$$@\, s = 0 \quad t = 0$$

$$0 = C$$

$$s(200+4t) = 20(200+2t^2)$$

$$V = 200 + 4t$$

$$240 = 200 + 4t$$

$T = 10$ min
$$s(200+4(10)) = 20(200+2t(10)^2)$$
$$s = 183.33N$$

$@\, t = 25$
$$s(200+4(25)) = 20(200+2t(25)^2)$$
$$s = 416N$$

16.   *A tank contains 400 litres of brine. Twelve litres of brine, each containing 2.5 N of dissolved salt, enter the tank per minute, and the mixture, assumed uniform, leaves at the rate of 8 litres per min. If the concentration is to be 2 N/ litre at the end of one hour, how many newtons of salt should there be present in the tank originally?*

Solution:

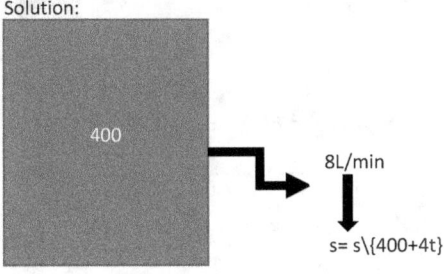

400

8L/min

s= s\{400+4t}

$$\frac{ds}{dt} = 30 - \frac{5(8)}{400 + 4t}$$

$$\theta = e^{\int \frac{8}{400+4t}}$$

$$\theta = e^{2\ln\left|\frac{8}{400+4t}\right|}$$

$$\theta = (400 + 4t)^2$$

$$S\left(\frac{400 + 4t}{4}\right)^2 = \frac{30}{3}\int \frac{(400 + 4t)^2}{4}\, d4$$

$$S(100 + t)^2 = 10\int (100 + t)^2 + C$$

$$@\, s =?\ \ t = 60$$

$$v = 400\big(4(60)\big) = 640L$$

$$s = 640\left(\frac{2N}{L}\right) = 1280N$$

$$1280(160)^2 = 10(160)^3 + C$$

$$C = -8192000$$

$$@\, t = 0\ \ s =?$$

$$s = \frac{10(100)^3 - 8192000}{100^3}$$

$$s = 180.8\ N$$

17. **Tank A initially contains 200 liters of brine containing 225 N of salt. Eight liters of fre water per minute enter A and the mixture, assumed uniform, passes from A to B, initially contain 200 liters of fresh water, at 8 liters per minute. The resulting mixture, also kept uniform, leave at the rate of 8 liters/min. Find the amount of salt in tank B after one hour.**

Solution:

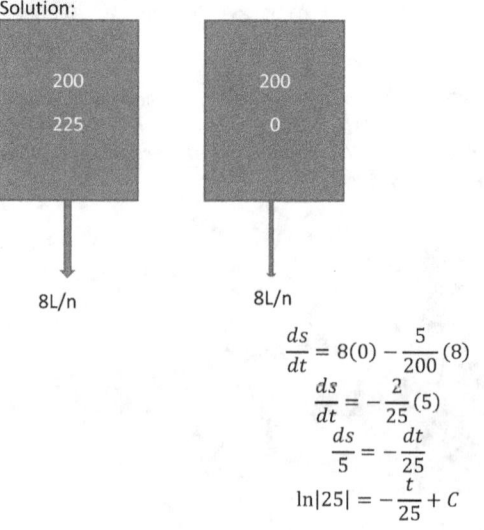

200
225

200
0

8L/n

8L/n

$$\frac{ds}{dt} = 8(0) - \frac{5}{200}(8)$$

$$\frac{ds}{dt} = -\frac{2}{25}(5)$$

$$\frac{ds}{5} = -\frac{dt}{25}$$

$$\ln|25| = -\frac{t}{25} + C$$

$$s = ce^{-\frac{t}{25}} \qquad @ \ t = 0$$
$$s = 22t \qquad\qquad 225 = C$$

$$S_A = 225e^{-\frac{t}{25}}$$
$$S = 225e^{-\frac{60}{25}}$$
$$S = 20.41 \ N$$
$$C = \frac{20.41}{200} = 0.102 \ \frac{N}{L}$$

$$\frac{ds}{dt} = \frac{S_A}{200} - \frac{S_B}{200}$$
$$\frac{ds_B}{dt} = \left(\frac{225e^{-\frac{t}{25}}}{200} - \frac{S_B}{200}\right)8$$
$$\frac{ds_B}{dt} = 9e^{-\frac{t}{25}} - \frac{S_B}{25}$$
$$\theta = e^{\int \frac{1}{25t}} \quad = e^{\frac{t}{25}}$$
$$\int \left(e^{\frac{t}{25}}\right) = 9 \int e^{\frac{t}{25}} \cdot e^{\frac{t}{25}} dt + C$$
$$\int \left(e^{\frac{t}{25}}\right) = 9 \int dt + C$$
$$\int \left(e^{\frac{t}{25}}\right) = 9t + C$$
$$@ \ t = 0 \quad S = 0 \quad C = 0$$
$$S = \frac{9(60)}{e^{\frac{60}{25}}}$$
$$S_B = 48.98N$$

18. **Tank A contains 400 liters of brine holding 225 N of dissolved salt, and tank B contains 400 liters of fresh water. If brine runs out of A into B at the rate of 12 liters/min while the mixture, kept thoroughly stirred, is pumped back from B to A at the same rate, when will A contain twice as much salt as B?**

Solution:

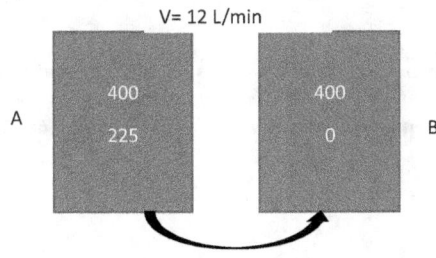

$$\frac{ds_A}{dt} = \frac{ds_A}{400}(12) - \frac{s_A(12)}{400}$$

$$\frac{ds_A}{ds_A} * \frac{ds_B}{dt} = \frac{s_A(12)}{400} - \frac{s_B(12)}{400}$$

$$\frac{ds_A}{dt} = \frac{ds_B}{dt}$$

$$\frac{ds_B}{ds_A}\left(\frac{3s_B}{100} - \frac{3s_A}{100}\right) = \left(\frac{3s_A}{100} - \frac{3s_B}{100}\right)$$

$$ds_B = -ds_A$$

$$s_B = -s_A + C$$

$$s_B = 0 \quad s_A = 225 \quad C = 225$$

$$s_B = -s_A + 225$$

$$\frac{ds_A}{dt} = \frac{(-s_A + 225)3}{100} - -\frac{3s_A}{100}$$

$$\frac{ds_A}{dt} = \frac{(-s_A + 225)3}{100} - -\frac{3s_A}{100}$$

$$\frac{ds_A}{dt} = \frac{(225)3}{100} - -\frac{6s_A}{100}$$

$$\frac{ds_A}{225(3) - 6s_A} = \frac{dt}{100}$$

$$-\frac{1}{6}|225(3) - 6s_A| = \frac{t}{100} + C$$

$$225(3) - 6s_A = Ce^{\frac{6t}{100}}$$

$$@ t = 0 \quad s_A = 225 \quad C - -675$$

$$225(3) - 6s_A = -675e^{-\frac{6t}{100}}$$

$$225(3) - 6(150) = -675e^{-\frac{6t}{100}}$$

$$-\frac{225}{-675} = e^{-\frac{6t}{100}}$$

$$\ln|0.333| = -\frac{6t}{100}$$

$$t = \frac{\ln|0.333|}{-\frac{6}{100}}$$

$$t = 18.3 \text{ min}$$

19.   1A tank with a horizontal sectional area constant at 10m2 and 4 m high contains water *
a depth of 3.5 m. The tank has a circular orifice 5 cm in diameter and is located at its side 0.5 *
above the bottom. If the coefficient of discharge of the orifice is 0.60, find the duration of flo*
through the orifice.

Solution:

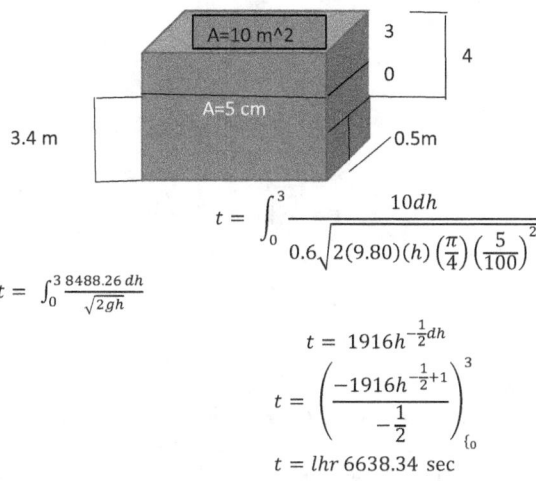

A=10 m^2

3
0
4

A=5 cm

3.4 m

0.5m

$$t = \int_0^3 \frac{10dh}{0.6\sqrt{2(9.80)(h)\left(\frac{\pi}{4}\right)\left(\frac{5}{100}\right)^2}}$$

$$t = \int_0^3 \frac{8488.26\, dh}{\sqrt{2gh}}$$

$$t = 1916h^{-\frac{1}{2}}dh$$

$$t = \left(\frac{-1916h^{-\frac{1}{2}+1}}{-\frac{1}{2}}\right)^3_{\{0}$$

$$t = 1hr\ 6638.34\ \text{sec}$$

$$or\ 1hr\ 50\ \min\quad and\ 38\ \sec$$

20.   *A tank in the shape of an inverted cone has a base diameter of 1.5 m and a height of 2 m. A 4-cm orifice (C= 0.60) is located at the apex. Determine the time required to empty the tank assuming that it is initially full of water.*

Solution:

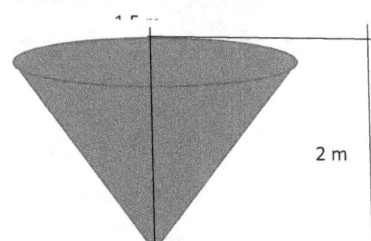

2 m

4 cm

$$A = \pi r^2$$
$$\frac{2}{0.75} = \frac{h}{r}$$

$$h = \frac{8}{3}r$$

15

$$t = \int_0^2 \frac{\pi \left(\frac{3}{8}h\right)^2 dh}{\frac{\pi}{4}\left(\frac{8}{100}\right)^2 (0.6)}$$

$$t = \int_0^2 132.21 \, h^{\frac{3}{2}}dh$$

$$t = \int_0^2 \left(\frac{132.21 \, h^{\frac{5}{2}}dh}{5}\right)_0^2$$

$$t = \int_0^2 \left(\frac{4(2)132.21 \, h^{\frac{5}{2}}dh}{5}\right)_0^2$$

$$t = 1200 \ sec \qquad or \ 20 \, mm$$

21. *A tank in the shape of an inverted frustum of a cone has an upper base diameter of 2 m a* *a lower base diameter of 1 m. A square orifice 5 cm x 5 cm (C=0.61) is located at the lower bas* *If the tank is 3 m high and initially full of water, find the time required for the water surface to dr* *2 m.*

Solution:

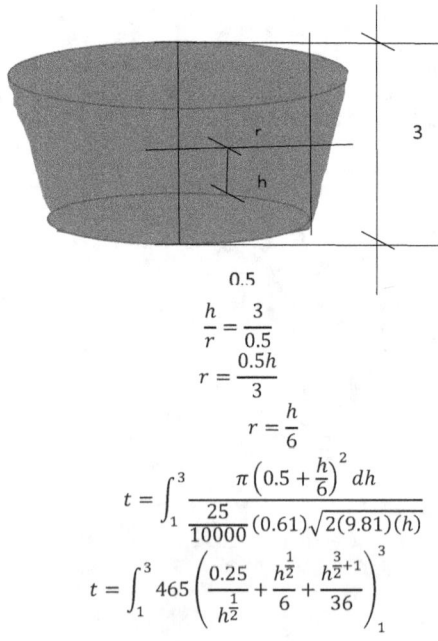

$$\frac{h}{r} = \frac{3}{0.5}$$

$$r = \frac{0.5h}{3}$$

$$r = \frac{h}{6}$$

$$t = \int_1^3 \frac{\pi\left(0.5 + \frac{h}{6}\right)^2 dh}{\frac{25}{10000}(0.61)\sqrt{2(9.81)(h)}}$$

$$t = \int_1^3 465 \left(\frac{0.25}{h^{\frac{1}{2}}} + \frac{h^{\frac{1}{2}}}{6} + \frac{h^{\frac{3}{2}+1}}{36}\right)^3_1$$

$$= 465\left(\frac{0.25\,h^{-\frac{1}{2}+1}}{-\frac{1}{2}+1}+\frac{h^{\frac{1}{2}+1}}{6\left(\frac{1}{2}+1\right)}+\frac{h^{\frac{3}{2}+1}}{36^{\frac{3}{2}+1}}\right)^{3}_{1}$$

$$t = 462 \text{ sec}$$

22. *A tank in the shape of a hemispherical shell has a diameter of 2 m. Water escapes through a circular orifice 10 cm in diameter (C 0.60) and located at the lowest point on the tank. Determine the time required to empty the tank assuming that it contains water amounting to half the volume of the tank.*

Solution:

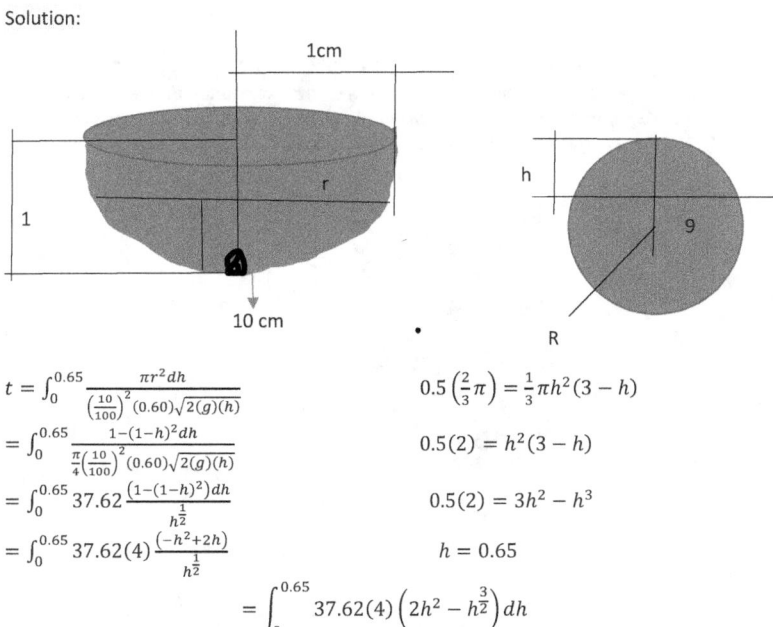

$$t = \int_0^{0.65} \frac{\pi r^2\,dh}{\left(\frac{10}{100}\right)^2(0.60)\sqrt{2(g)(h)}}$$

$$= \int_0^{0.65} \frac{1-(1-h)^2\,dh}{\frac{\pi}{4}\left(\frac{10}{100}\right)^2(0.60)\sqrt{2(g)(h)}}$$

$$= \int_0^{0.65} 37.62\,\frac{\left(1-(1-h)^2\right)dh}{h^{\frac{1}{2}}}$$

$$= \int_0^{0.65} 37.62(4)\,\frac{(-h^2+2h)}{h^{\frac{1}{2}}}$$

$$= \int_0^{0.65} 37.62(4)\left(2h^2 - h^{\frac{3}{2}}\right)dh$$

$$0.5\left(\frac{2}{3}\pi\right) = \frac{1}{3}\pi h^2(3-h)$$

$$0.5(2) = h^2(3-h)$$

$$0.5(2) = 3h^2 - h^3$$

$$h = 0.65$$

$$= \left(37.62(4) \left(\frac{2h^2}{\frac{1}{2}+1} - \frac{h^{\frac{3}{2}}}{\frac{3}{2}+1} \right) \right)_0^{0.65} \qquad 1^2 = r^2 + (1-h)^2$$

$$= \left(150.48 \left(2h^{\frac{3}{2}} \left(\frac{2}{3} \right) - h^{\frac{5}{2}} \left(\frac{2}{5} \right) \right) \right)_0^{0.65} \qquad r^2 = 1^2 - (1-h)^2$$

$$= 150.48\,(0.56)$$
$$t = 85 \text{ sec}$$

23. **A swimming pool has a gradually sloping bottom. It has a depth of 1 m at one end and depth of 3 m at the other end. The horizontal sectional shape of the pool is a rec- tangle 50 m x m. Two short tubes each 15 cm in diameter (C= 0.80) are located at the lowest level of the pool the pool is initially full of water, find the time to empty the pool through the tubes.**

Solution:

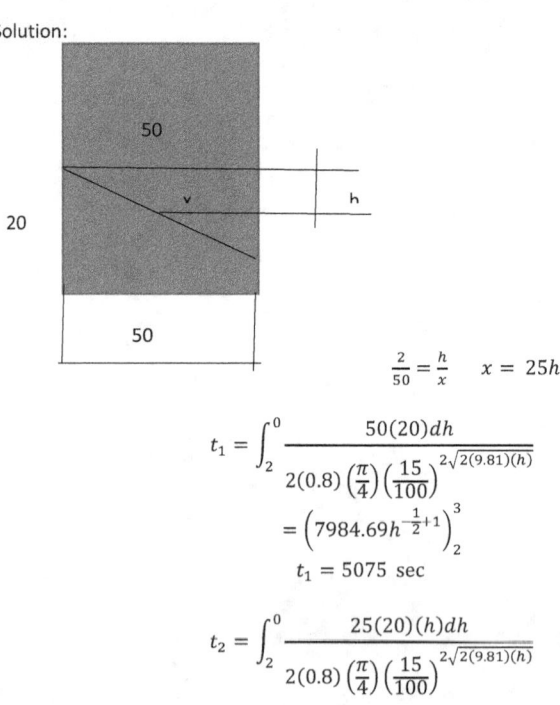

$$\frac{2}{50} = \frac{h}{x} \qquad x = 25h$$

$$t_1 = \int_2^0 \frac{50(20)dh}{2(0.8)\left(\frac{\pi}{4}\right)\left(\frac{15}{100}\right)^2 \sqrt{2(9.81)(h)}}$$

$$= \left(7984.69 h^{-\frac{1}{2}+1} \right)_2^3$$

$$t_1 = 5075 \text{ sec}$$

$$t_2 = \int_2^0 \frac{25(20)(h)dh}{2(0.8)\left(\frac{\pi}{4}\right)\left(\frac{15}{100}\right)^2 \sqrt{2(9.81)(h)}}$$

$$= \left(\frac{3992.35 \, h^{\frac{1}{2}} dh}{\frac{1}{2} + 1} \right)^2_0$$

$$t_2 = 7528 \text{ sec}$$

$$t_T = 5075 + 7528 \text{ sec}$$
$$t_T = 126033 \text{ sec}$$

24. **A coil of inductance 1 henry and resistance 10 ohms is connected in series with an e.m.f. of Eo sin 10t volts. When t=0, the current is zero. If I=5 amp, when t =0.1 sec, what must be the value of Eo?**

Solution:

$$EL = L \frac{dl}{dt}$$
$$ER = Rl$$
$$EL = \frac{Q}{C}$$
$$I = \frac{dQ}{dt}$$
$$L \frac{dl}{dt} + Rl + \frac{Q}{C} = E$$
$$L \frac{d^2Q}{dt^2} + R \frac{dQ}{dt} + \frac{Q}{C} = E$$
$$E - EL - ER - EC = 0$$
$$L = 1 \, henry \quad R = 10 \, \Omega$$
$$E \sin\theta \, 10t$$
$$\frac{d^2Q}{dt^2} + \frac{10dQ}{dt} = E \sin \quad 10t$$
$$\frac{dl}{dt} + 10 = E \sin \quad 10t$$
$$Q = e^{\int 10 \, dt} \qquad = E \sin \quad 10t$$
$$D = E_0 \sin \quad 10t$$

$$I \, E_0 e^{10t} = E \cdot \int$$

$$I \, E_0 e^{10t} = E \cdot \int \sin 1 \, 0t \cdot e^{10t} dt + C$$
$$v = \sin 1 \, 0t \quad d = e^{10t}$$

$$dv = 10 \cos 1\,0t \quad v = \frac{e^{10t}}{10}$$

$$\frac{\sin 1\,0\,e^{10t}}{10} - \int \cos 1\,0t\, e^{10t}$$

$$v = \cos 1\,0t \quad dv = e^{10t}$$

$$dv = -\sin 1\,0t\,(10) \quad v = \frac{e^{10t}}{10}$$

$$I\,e^{10t} = E_0 \left(\frac{\sin 1\,0t\, e^{10t}}{10} - \frac{\cos 1\,0t\, e^{10t}}{10} \right) = \int -\frac{10\sin t\ e^{10t}}{10}$$

$$I\,e^{10t} = \frac{E_0}{10} \left(\frac{\sin 1\,0t\, e^{10t} - \cos 1\,0t\, e^{10t}}{2} \right) + C$$

$$@\,t = 0 \quad I = 0$$

$$0 = E_0 \left(\frac{0-1}{2} \right) + C$$

$$c = \frac{E_0}{20}$$

$$\frac{I\,e^{10t}}{\frac{e^{10t}}{20}(\sin 10t - \cos 10t) + \frac{1}{20}} = E_0 \qquad @\,t = 0.1 \quad i = 5$$

$$E_0 = 149.5\,Volts$$

25.   An inductance of I henry and a resistance of 2 ohms are connected in series with an e.m. of E et volts. No cur. rent is flowing initially. (a) If the current 1 10 amp. when t1 sec, how muc must E be? (b) If E=50 volts, when will the current be 5 amp?

Solution:

$$L = 1 \quad R = 2 \quad T = 0 \quad I = 0$$

$$\frac{dI}{dt} + 21 = E e^{-t}$$

$$\theta = e^{\int 2dt} = e^{2t}$$

$$I e^{2t} = \int E e^{-t}(e^{2t})dt + C$$

$$I e^{2t} = E e^{t}\,dt + C$$

$$@\,t = 0 \quad I = 0 \qquad C = -e$$

$$I e^{2t} = E e^{t} - E$$

$$@\,I = 10 \quad t = 1$$

$$E = \frac{I\,e^{2t}}{e^{t} - 1}$$

$$E = 43V$$

$$50 = \frac{5e^{2t}}{e^{t} - 1}$$

$$50e^{t} - 50 = 5e^{2t}$$

$$50e^t + 50 - 5e^{2t} = 0$$
$$e^t = 1 - 12 \quad e^t = 8.87$$
$$t = \ln|1.12| \quad t = \ln|8.87|$$
$$t = 0.113 \text{ sec} \quad t = 2,18 \text{ sec}$$

26.   **An inductance L Henries and resistance R ohms are connected in series with an e.m.f. of E eat volts where E and a are positive constants. Initially, the current I is zero. Find an expression for 1 as a function of t, and determine at what time the current reaches its maximum value.**

Solution:

$$L\frac{dI}{dt} + RI = Ee^{9t}$$

$$T = 0 \quad I = 0 \quad I = F(t) \quad Imax =?$$

$$\frac{dI}{dt} + \frac{RI}{L} = Ee^{-9t}$$

$$\theta = e^{\int \frac{R}{L} - 9t} \quad = e^{\frac{RL}{L}}$$

$$Q = \frac{E}{L}e^{-9t}$$

$$I\,e^{\frac{Rt}{L}} = \frac{E}{L}\int e^{et\left(\frac{R}{L}\right)}\,dt + C$$

$$I\,e^{\frac{Rt}{L}} = \frac{\frac{E}{L}\left(e^{et\left(\frac{R}{L}-9\right)}\right)}{\frac{R}{L} - 9} + C$$

$$I\,e^{\frac{Rt}{L}} = -\frac{E}{R-L9}\left(e^{t\left(R-\frac{L9}{L}\right)}\right) + C$$

$$@\ I = 0 \quad t = 0 \quad 0 = -\frac{E}{R-L9} = C$$

$$I = \frac{\frac{E}{R-L9}\left(e^{t\left(\frac{R-La}{L}\right)}-1\right)}{e^{\frac{Rt}{L}}}$$

$$I = \frac{E}{R-L9}\left(e^{-\frac{L9t}{L}} - e^{-\frac{Rt}{L}}\right)$$

$$I = \frac{E}{R-L9}\left(e^{-\frac{9t}{L}} - e^{-\frac{Rt}{L}}\right)$$

$$0 = \frac{dI}{dt} = \frac{E}{R-L9}\left(e^{-\frac{9t}{L}} - \frac{R}{L}e^{-\frac{Rt}{L}}\right)$$

$$9e^{-9t} = \frac{R}{L}e^{-\frac{Rt}{L}}$$

$$\frac{L9}{R} = e^{-\frac{Rt}{L}} + 9t$$

$$\frac{L\ln\left|\frac{L9}{R}\right|}{0L - R} = t$$

27. *A pipe 20 cm in diameter contains steam at 100 °C and is covered with a certain insulati*
5 cm thick. The outside temperature is kept at 40 °C. By how much should the thickness
insulation be increased in order that the rate of heat loss shall be decreased 20%?

Solution:

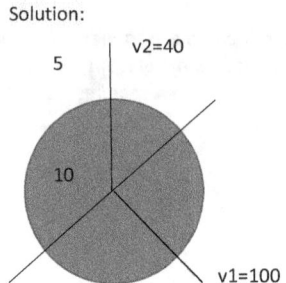

$$9 = -kA \frac{dv}{dx}$$

$$9 = \frac{2\pi kL(v_1 - v_2)}{\ln\left|\frac{x_2}{x_1}\right|}$$

$$9 = \frac{2\pi kL(100 - 40)}{\ln\left|\frac{15}{10}\right|}$$

$$9 = \frac{2\pi kL(100-40)}{\ln\left|\frac{15}{10}\right|}(0.8) = \frac{2\pi kL(100-40)}{\ln\left|\frac{x}{10}\right|}$$

$$1.97 = \frac{1}{\ln\left|\frac{x}{10}\right|}$$

$$\ln\left|\frac{x}{10}\right| = \frac{1}{1.97}$$

$$\frac{x}{10} = e^{\frac{1}{1.97}}$$

$$x = 10(1.66)$$

$$x = 16.61 \; cm$$

$$\Delta x = 15 - 16.61 \; cm$$

$$\Delta x = 1.61 \; cm$$

28. *A steam pipe of radius 3 cm and at 100°C is wrapped with a 1 cm layer of insulation*
thermal conductivity 0.0003 cal/cm.deg.sec and then that layer is wrapped with a 2 cm layer
insulation of thermal conductivity 0.0002 cal/cm.deg sec At what temperature must the out- si
surface be maintained in order that 0.008 cal/sec will flow from each square em of pipe surface

Solution:

$$0.008 = k \frac{dv}{dx}$$

$$\frac{0.008}{\pi(6)} = 0.0003 \frac{dv}{dx}$$

$$1.41(4) = v - 95.77$$

$$0.1508L \frac{cal}{sec}$$

$$0.008(\pi)(6)(L) = \frac{2\pi(0.0003)(L)(100 - v_2)}{\ln \left|\frac{4}{3}\right|}$$

$$v_2 - 100 = 23.015$$
$$v_2 = 76.99\ ^oC$$

$$\frac{2\pi(0.0003)(L)(100 - v_2)}{\ln \left|\frac{4}{3}\right|} = \frac{2\pi(0.0003)(L)(v_2 - v_3)}{\ln \left|\frac{6}{4}\right|}$$

$$2.11(100 - v_2) = v_2 - v_3$$
$$2.11(100 - 76.94) = 76.94 - v_3$$

$$v_3 = 28.44^oC$$

29.   *A pipe 10 cm in diameter, contains steam at 100°C. It is to be covered with two coats of insulating material each 2.5 cm thick; the inner with K= 0.0006 cal/cm.deg.sec and the outer with K = 0.00017 cal/cm.deg.sec. If the outside surface temperature is 30°C, find the heat loss per hour from a meter length of pipe."*

Solution:

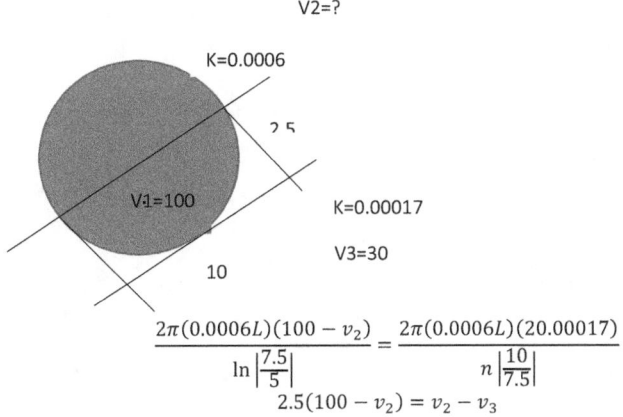

$$\frac{2\pi(0.0006L)(100 - v_2)}{\ln \left|\frac{7.5}{5}\right|} = \frac{2\pi(0.0006L)(20.00017)}{n \left|\frac{10}{7.5}\right|}$$

$$2.5(100 - v_2) = v_2 - v_3$$

23

$$v_3 = 30$$
$$2.5(100 - v_2) = v_2 - 30$$
$$v_2 = 80$$

$$\frac{9}{L} = \frac{2\pi(0.0006)(100 - 80)}{\ln\left|\frac{7.5}{5}\right|}$$

$$0.18595 \; \frac{cal}{sec \quad \cdot cm} \cdot 3600$$

$$667.438 \; \frac{cal}{hr \cdot cm} \cdot \frac{1000}{1m}$$

$$667.438 \; \frac{cal}{hr \cdot m}$$